GeoPlanet: Earth and Planetary Sciences

More information about this series at http://www.springer.com/series/8821

Maria Jeleńska · Leszek Łęczyński
Tadeusz Ossowski
Editors

Magnetometry in Environmental Sciences

Studying Environmental Structure Changes and Environmental Pollution

Springer

Editors
Maria Jeleńska
Institute of Geophysics
Polish Academy of Sciences
Warsaw
Poland

Tadeusz Ossowski
Faculty of Chemistry
University of Gdańsk
Gdańsk
Poland

Leszek Łęczyński
Institute of Oceanography
University of Gdańsk
Gdynia
Poland

The GeoPlanet: Earth and Planetary Sciences Book Series is in part a continuation of Monographic Volumes of Publications of the Institute of Geophysics, Polish Academy of Sciences, the journal published since 1962 (http://pub.igf.edu.pl/index.php).

ISSN 2190-5193 ISSN 2190-5207 (electronic)
GeoPlanet: Earth and Planetary Sciences
ISBN 978-3-319-60212-7 ISBN 978-3-319-60213-4 (eBook)
DOI 10.1007/978-3-319-60213-4

Library of Congress Control Number: 2017943175

Printed on acid-free paper

This Springer imprint is published by Springer Nature
The registered company is Springer International Publishing AG
The registered company address is: Gewerbestrasse 11, 6330 Cham, Switzerland

Series Editors

Managing Editor

Anna Dziembowska
Institute of Geophysics, Polish Academy of Sciences

Advisory Board

Kon-Kee Liu
Institute of Hydrological and Oceanic Sciences
National Central University Jhongli
Jhongli, Taiwan

Teresa Madeyska
Research Centre in Warsaw
Institute of Geological Sciences
Warszawa, Poland

Stanisław Massel
Institute of Oceanology
Polish Academy of Sciences
Sopot, Poland

Antonio Meloni
Instituto Nazionale di Geofisica
Rome, Italy

Evangelos Papathanassiou
Hellenic Centre for Marine Research
Anavissos, Greece

Kaja Pietsch
AGH University of Science and Technology
Kraków, Poland

Dušan Plašienka
Prírodovedecká fakulta, UK
Univerzita Komenského
Bratislava, Slovakia

Barbara Popielawska
Space Research Centre
Polish Academy of Sciences
Warszawa, Poland

Tilman Spohn
Deutsches Zentrum für Luftund
Raumfahrt in der Helmholtz
Gemeinschaft
Institut für Planetenforschung
Berlin, Germany

Krzysztof Stasiewicz
Swedish Institute of Space Physics
Uppsala, Sweden

Ewa Szuszkiewicz
Department of Astronomy
and Astrophysics
University of Szczecin
Szczecin, Poland

Roman Teisseyre
Department of Theoretical Geophysics
Institute of Geophysics
Polish Academy of Sciences
Warszawa, Poland

Jacek Tronczynski
Laboratory of Biogeochemistry
of Organic Contaminants
IFREMER DCN_BE
Nantes, France

Steve Wallis
School of the Built Environment
Heriot-Watt University
Riccarton, Edinburgh
Scotland, UK

Wacław M. Zuberek
Department of Applied Geology
University of Silesia
Sosnowiec, Poland

Piotr Życki
Nicolaus Copernicus Astronomical
Centre
Polish Academy of Sciences
Warszawa, Poland

Contents

Introduction

Environmental Magnetism investigates magnetic properties of the medium in relation to processes which formed the environment, as magnetic properties are controlled by minerals containing iron and are very sensitive to geological events, biological processes, climate and man activity.

Magnetometry is a set of magnetic methods used to solve different problems connected with the environment, such as:

Pollution of air, water and soil
Pedogenic processes and soil structure
Magnetic properties of lake, river and marine sediments in relation to sedimentation processes.

Magnetism is a common feature of matter, as the electrons rotating on atom orbits make orbital and spin movements and generate the magnetic field. They behave as small magnets and react on the external magnetic field. All substances react to the presence of magnetic field and create an induced magnetization. The relationship between the external magnetic field and the induced magnetization is called magnetic susceptibility and is defined as:

$$K = M/H,$$

where M is magnetization and H is magnetic field.

There are three categories of this relation.

Diamagnetic Substances are magnetized in the direction opposite to the magnetic field direction. Diamagnetic susceptibility is negative and independent of temperature. The effect is very weak and after removal of field diamagnetic magnetization goes to zero. This is the property of all substances.

Paramagnetic substances have unpaired electron spins which create magnetization, essentially randomly oriented. The external magnetic field orders the spins and creates bulk magnetization. The paramagnetic susceptibility is positive, larger than the diamagnetic one and inversely proportional to temperature.

Ferromagnetic substances (sensu lato) have the ability to keep magnetization in the absence of magnetic field. This is the so-called remanent magnetization or spontaneous magnetization. In these substances, spins are aligned parallel (ferromagnetism) or anti-parallel (antiferromagnetism or ferrimagnetism) due to strong interactions between neighboring spins in certain crystal lattices. Ferromagnetic susceptibility is positive, sometimes very strong, nonlinearly dependent on the magnetic field.

The alignment of spins can be parallel in ferromagnetic substances (sensu stricto) or perfectly anti-parallel as in antiferromagnetism. When anti-parallel ordering is not perfect or not completely compensated because of crystal defects, weak net moment occurs. In ferrimagnetism, spins are ordering perfectly anti-parallel but they have different moments (Fig. 1).

With increasing temperature, the spin alignment is destroyed by thermal fluctuations and ferromagnetic substances become weaker and, at temperature characteristic of each substance, they become paramagnetic. This is the Curie temperature for ferro- and ferrimagnetism and the Neel temperature for antiferromagnetism.

The Curie temperature is used for identification of magnetic minerals. It is obtained from thermal decay curves of saturation magnetization M(T), saturation remanence SIRM(T) or susceptibility κ(T).

M(T) and κ(T) curves give the value of Curie temperature, whereas SIRM(T) curve gives the unblocking temperature which is close to the Curie temperature.

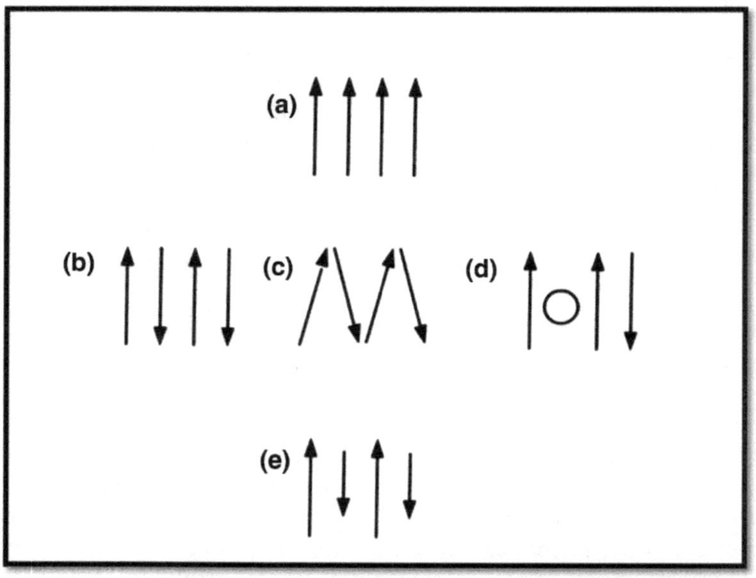

Fig. 1 Spin ordering in ferromagnetism (sensu lato). **a** ferromagnetism; **b** antiferromagnetism; **c** spin-canted antiferromagnetism; **d** defect antiferromagnetism; **e** ferrimagnetism

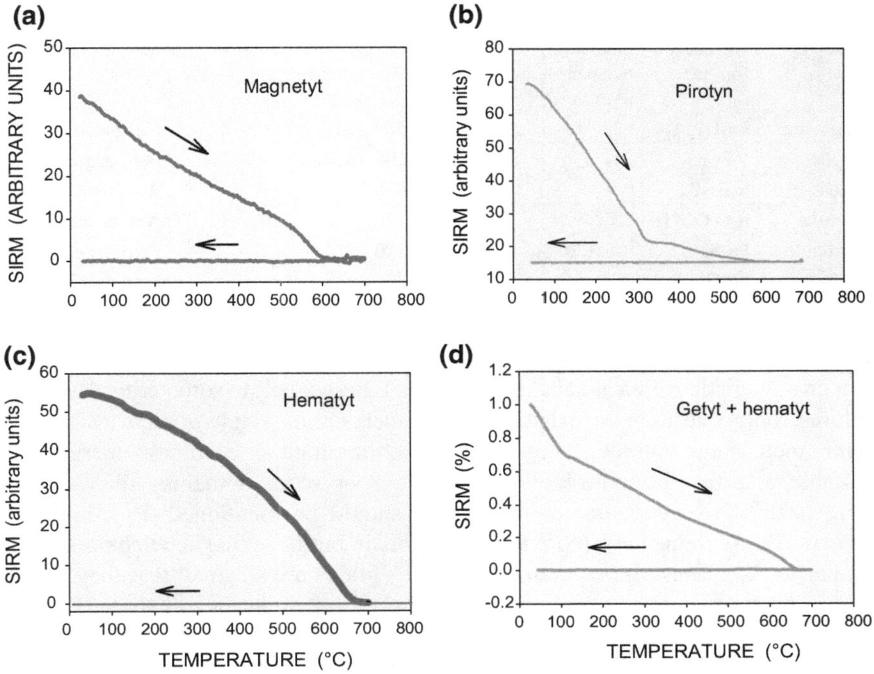

Fig. 2 Examples of SIRM(T) curves for **a** magnetite; **b** pyrrhotite; **c** hematite; **d** goethite and hematite

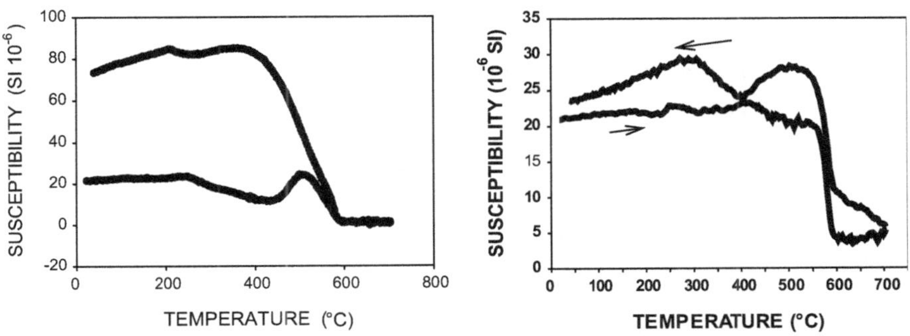

Fig. 3 Examples of κ(T) curves for soils and dust

Examples of SIRM(T) curves and κ(T) curves for some minerals are shown in Figs. 2 and 3.

Main magnetic minerals occurring in natural environment are presented in Table 1.

The alignment of unpaired electronic spins occurs over a large area within the crystal, creating the so-called magnetic domains. The alignment of spins within the

Table 1 Main magnetic minerals occurring in natural environment

Mineral	Chemical formula	Saturation magnetization Ms (Am^2/kg)	Curie temperature (°C)	K (SI)	Magnetic ordering
Magnetite	Fe_3O_4	92	580	1	Ferrimagnetic
Maghemite	γ-Fe_2O_3	74	590–675	–	Ferrimagnetic
Hematite	α-Fe_2O_3	0.4	675	1.3×10^{-3}	Antiferromagnetic
Goethite	α-FeOOH	$10^{-3} - 1$	120	1.0×10^{-3}	Antiferromagnetic
Pyrrhotite	Fe_7S_8	$0.4 - 20$	320	1.0×10^{-3}	Ferrimagnetic
Greigite	Fe_3O_4	25	330–350	–	Ferrimagnetic

domain gives rise to spontaneous magnetization—the highest value of magnetization characteristic for each substance. In small grains, below some critical value of volume, only one domain exists. We call such grains single-domain (SD) ones. With increasing volume, single-domain configuration becomes energetically unstable and the grain broke into two, four or more domains—multi-domain (MD) state. Two particular domain states should be mentioned. Pseudo-single-domain (PSD) grains have more than one domain but have magnetic characteristics similar to SD grains. Superparamagnetic (SP) grains are so small that they are not able to retain magnetic remanence. SP grains are important in soil study. Figure 4 demonstrates different domain states.

The response of ferromagnetic substances (sensu lato) to external magnetic field action is characterized by hysteresis phenomena (Fig. 5). Magnetization of ferromagnetic substances does not increase linearly with increasing field. At a field of intensity characteristic for particular substance, the magnetization reaches saturation (M_s). After removal of field, magnetization does not fall to zero but decreases to the

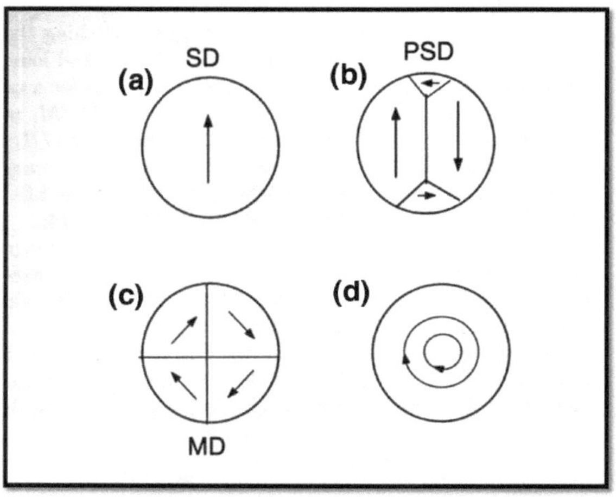

Fig. 4 Different domain states. **a** single domain; **b** pseudo-single-domain; **c** multi-domain; **d** superparamagnetic state

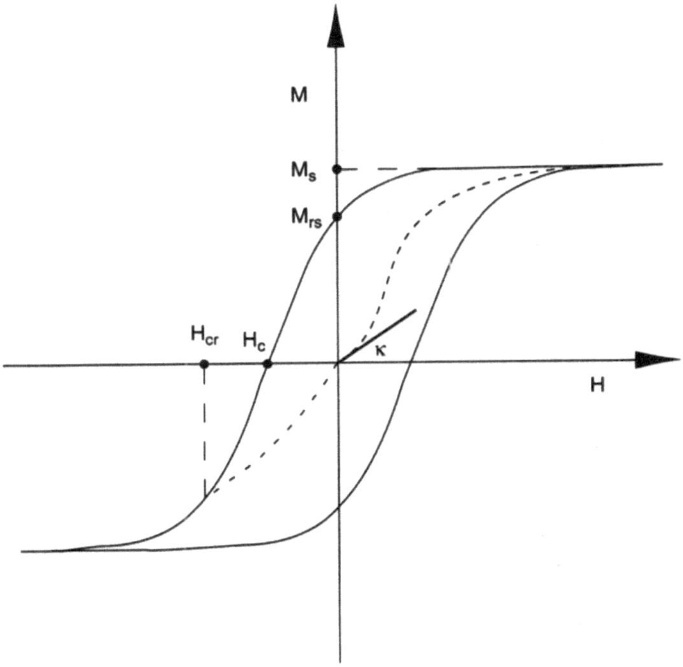

Fig. 5 Hysteresis loop. Ms—saturation magnetization; Mrs—saturation remanence; Hc—coercivity; Hcr—coercivity of remanence; κ—low field susceptibility

value called remanence or remanent magnetization (M_{rs}). Magnetization reaches zero value after applying field of characteristic value called coercive force or coercivity (H_c) in opposite direction. The opposite field required for reducing remanence to zero is called coercivity of remanence (H_{cr}). The slope of initial magnetization curve at the beginning of magnetization process in its linear and reversible distance is defined as low field susceptibility κ—a very important magnetic parameter in environmental magnetism study. Figure 6 shows different responses of substances on application of magnetic field and different magnetization processes of diamagnetic, paramagnetic and ferromagnetic substances of different domain state.

The following magnetic parameters are used in environmental magnetism:

κ or χ is the low field magnetic susceptibility calculated for volume unit or mass unit

χ_{fd} is the dependence of susceptibility on frequency of magnetic field defined as $\chi_{fd} = \chi_{lf} - \chi_{hf}$, where χ_{lf} is the susceptibility measured in low frequency field and χ_{hf} is the susceptibility measured in high frequency field. Often $\chi_{fd}(\%)$ is used in relation to the χ_{lf} value.

ARM is the anhysteretic remanence obtained by acquiring remanence during applying alternating field which reduces from 100 mT amplitude to 0 in the

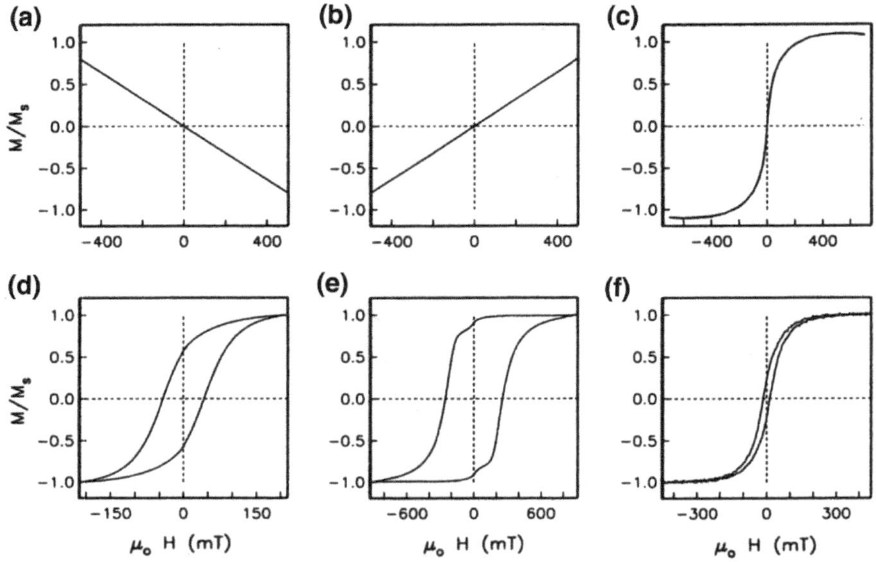

Fig. 6 Examples of magnetization processes of diamagnetic **a**, paramagnetic **b** and ferromagnetic substances of different domain state: SP **c**; SD **d, e**; PSD **f**

presence of constant bias field of 100 μT. Usually it is expressed as χ_{ARM} by dividing ARM by constant field value.

Soft IRM = (SIRM − $IRM_{-0.3T}$)/2 often expressed as the S-ratio S = $IRM_{-0.3T}$ /SIRM and hard IRM is HIRM = (SIRM + $IRM_{-0.3T}$)/2, where SIRM is the saturation remanence acquired in 1T field and $IRM_{-0.3T}$ is the IRM value remained after remagnetization SIRM by −03.T field. These parameters are sensitive to changes of magnetic mineralogy.

All magnetic parameters characterize the kind of magnetic minerals present in the measured substance.

Susceptibility, saturation magnetization and saturation remanence and anhysteretic remanence depend also on the content of magnetic minerals. They are concentration-dependent parameters. Except for saturation magnetization, they depend not only on concentration, but also on size of magnetic particles. Susceptibility is lower for SD particles than for PSD and MD particles and is very high for SP particles. ARM is the highest for SD particles.

Coercivity and coercivity of remanence depend on size and shape of magnetic particles. They are not dependent on concentration of magnetic minerals. Small and elongated particles have the highest coercivity and coercivity of remanence values.

Frequency dependence of susceptibility χ_{fd} is the highest for particle size on the border of SP and SD grains. High value of this parameter is characteristic for magnetic particles of pedogenic origin.

Besides these parameters, ratio parameters are in common use for magnetic characteristics of environment. Hysteresis parameters: M_s, M_{rs}, H_c and H_{cr} are presented on Day diagram as bi-plot of squareness M_{rs}/M_s versus H_{cr}/H_c. They are sensitive to the domain state: SP, Sd PSD and MD. Therefore, the Day plot is used for approximation of the domain state of ferromagnetic substance.

The ratio χ_{ARM}/χ_{fd} demonstrated as diagram is used for approximation of changes in grain size of homogeneous mineralogy.

ARM or χ_{ARM} in relation to SIRM reflects particle interactions, as the ARM acquisition is hampered by particle interaction, while SIRM acquisition is not.

All the above-mentioned parameters are demonstrated as maps, bi-plots and depth plots. They add numerous helpful pieces of information to geochemical, mineralogical and sedimentological data.

Maria Jeleńska

Magnetometric Assessment of Soil Contamination in the Vicinity of Selected Roads in Poland

Olga Rosowiecka and Jerzy Nawrocki

Abstract The goal of the paper is to describe the degree of applicability of magnetic susceptibility measurements for determination of lateral and vertical extension of soil pollution in the neighborhood of selected sections of roads in Poland. Magnetic susceptibility measurements were supported by petromagnetic, geochemical and SEM analyses. In the case of subordinate roads, increased magnetic susceptibility values appears only at the shoulder of the road, i.e., within 3 m from the edge of the road, while in the case of main roads an anomalous band usually reaches about 15 m. Increased values of magnetic susceptibility were observed up to a depth of 30 cm. Main magnetic carriers are Fe and Ti oxides and metallic iron, appearing as irregular grains and spherules. At some grains containing Fe and Ti, fragments of crystal walls were observed. Other grains appeared as elongated slivers. Fe and Ti oxides with remains of crystal walls are of natural, i.e., geogenic nature. Their concentration in the vicinity of the road results from degradation (elution, erosion) of gravel and sand foundation of the road. Irregular grains and metallic iron are coming from traffic (shards of car bodies, combustion products). Magnetic susceptibility carriers of soil samples taken from anomalous areas are associated with elements such as zinc, lead, sulfur and (to a less extent) nickel. Increase of magnetic susceptibility at the topsoil level is accompanied by increasing content of these elements. The zone up to 15 m away, and somewhere even up to 40 m away from the road edges should be excluded from cultivation.

Keywords Magnetic susceptibility · Soil contamination · Road vicinity

---

O. Rosowiecka (✉)
Polish Geological Institute, National Research Institute, Rakowiecka 4,
00-975 Warsaw, Poland
e-mail: olga.rosowiecka@pgi.gov.pl

J. Nawrocki
Department of Earth Sciences and Spatial Management,
Maria Curie-Skłodowska University, al. Kraśnicka 2cd, 20-718 Lublin, Poland
e-mail: jerzy.nawrocki@pgi.gov.pl

M. Jeleńska et al. (eds.), *Magnetometry in Environmental Sciences*,
GeoPlanet: Earth and Planetary Sciences, DOI 10.1007/978-3-319-60213-4_1

1 Introduction

Roads, or more precisely the car traffic, have been detected as a significant source of pollution of soils by heavy metals (e.g. Hofmann et al. 1999; Lu et al. 2005). Before unleaded petrol came into the common use, it had been found that lead content tends to occur together with the magnetic carriers (Beckwith et al. 1990; Matzka and Maher 1999). Beckwith et al. (1986, 1990) have observed that values of magnetic susceptibility are relatively high near the roads and they decrease rapidly outside of them, being a function of distance from the particular road. In a similar way, the total content of some elements like Fe, Pb, Zn and Cu was changed. They have concluded that car transport is the cause of increased values of magnetic susceptibility. Matzka and Maher (1999) have studied magnetic properties of birch leafs from single tree growing in the center of Norwich. Leafs from the road side were substantially more magnetized than leafs from the opposite side of trunk. It means that the roadside trees not only shade the roads but they capture a certain part of heavy metals coming from the road traffic as well. Studies of magnetic susceptibility as an indicator of magnitude of road pollution were also carried out by many other teams (e.g. Hoffmann et al. 1999; Kletetschka et al. 2003; Goddu et al. 2004; Bućko et al. 2011). Hoffmann et al. (1999) point out that magnetic susceptibility values near the roads are even 10 times higher than far away, and the abrasion products from the asphalt and from vehicle brake systems are the main source of magnetic pollution. Jordanova et al. (2014) have noticed that high reverse correlation between anhysteretic remanent magnetization and Pb content indicates larger magnetic grains as carriers of this element. Kim et al. (2007) have indicated that compounds of ferric oxides with carbon and sulphur were connected with traffic emission, whereas pure iron was emitted from car brakes.

The aim of this study was to determine the degree of usefulness of magnetic susceptibility field measurements for defining the degree and lateral extension of soil pollution near selected roads in Poland. In particular, the distribution of magnetic susceptibility value depending on distance from the roads, types of soils and locations of roads with reference to prevailing wind directions was examined. Studies of magnetic susceptibility and geochemical analysis of soils in vertical sections were expected to permit an assessment of the contamination dept range and its dependence on soil type. To define which elements coming from the road traffic are associated with the magnetic carriers, a comparison of magnetic susceptibility data with the results of geochemical studies was performed. The main advantage of field measurements of magnetic susceptibility is, however, a relatively fast acquisition of data and their relatively low cost.

2 Study Localities and Methods

Seventeen locations (sections of roads) were chosen to represent different road classes (motorways, trunk roads, primary roads), different soil types (chernozems, phaeosols, cambisols, alluvial soils, podzols), and different topography. Locations of our studies are shown in the topography and road net background (Fig. 1). The road map was drawn on the basis of the OSM (© authors OpenStreetMap). Sites are named according to road numbers, e.g., 47/1 section is the first section at primary road number 47. All road sections were divided into four geographical groups, i.e., mountain areas—Carpathian Mts. (sections: 47/1, 47/2, 47/3, 47/4, 957); mountain areas—Sudetes (sections: 33, 381, 386, 8/1); uplands (sections: 7/1, 7/2, 776, 812, A4) and lowlands (sections: 7/3, 7/4, 7/5). Color scales have been tailored to emphasize anomalous zones of each case study, because any classification of pollution areas depends on the background values that are different for particular types of soils.

For each location, a magnetic susceptibility (κ) map was created basing on field measurements. The measurements were performed using MS2 Bartington kappabridge connected to MS2D sensor. They were distributed along profiles perpendicular to the road axis. Average distance between profiles was of c.a. 10 m. A measuring step on the profile increases from 1 m at the road edge up to 5 m at the other end. 21 vertical soil profiles were collected to examine a vertical extension of pollution. They were sampled for laboratory measurements. The soil profiles were located both in the zone of increased κ, close to the road, and in the background zone, further away from the road. Thus, it is possible to compare a potentially polluted material with a relatively clean one.

Laboratory measurements have included the mass magnetic susceptibility (χ) measurements, the isothermal remanent magnetization (IRM) analyses, thermomagnetic analyses of magnetic susceptibility, geochemistry and scanning electron microscope (SEM) observations. Measurements of χ were performed in two frequencies: 0.47 kHz (χ_{lf}) and 4.7 kHz (χ_{hf}) by the means of MS2 Bartington kappabridge connected to MS2B sensor. The χ_{Fd} coefficient, resulting from χ_{lf} and χ_{hf}, measurements was used as a grain size indicator. The IRM was acquired using the Magnetic Measurements pulse magnetizer. Samples were magnetized in increasing fields of amplitude up to 2.5 T. Measurements of remanence were done with JR5 magnetometer. Thermoanalyses of magnetic susceptibility were performed using the MS2 unit connected to MS2WT sensor, within the temperature increasing up to 900 °C. Geochemical analyses were conducted by the Central Chemical Laboratory of Polish Geological Institute (PGI-NRI). All elements were detected using the ICP-AES method.

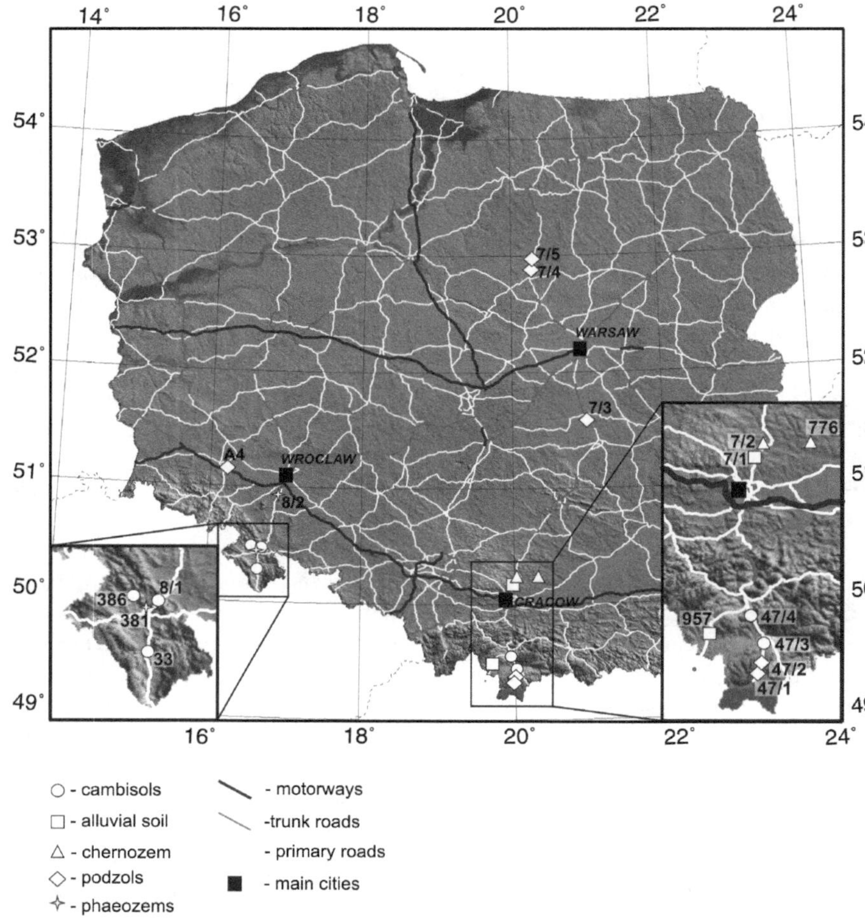

Fig. 1 Location of studied road sections on the background of map of roads in Poland

3 Results

3.1 *General Trends in Distribution of Magnetic Susceptibility on the Surface*

Figures 2, 3, 4 and 5 show maps of magnetic susceptibility of individual groups of the road sections. The slope direction is presented when it is higher than 1%. A soil type is also marked on the maps. A background magnetic susceptibility κ_{back} (κ of clean soil) depends strongly on it. Chernozem developed on loess (sections 7/2 and

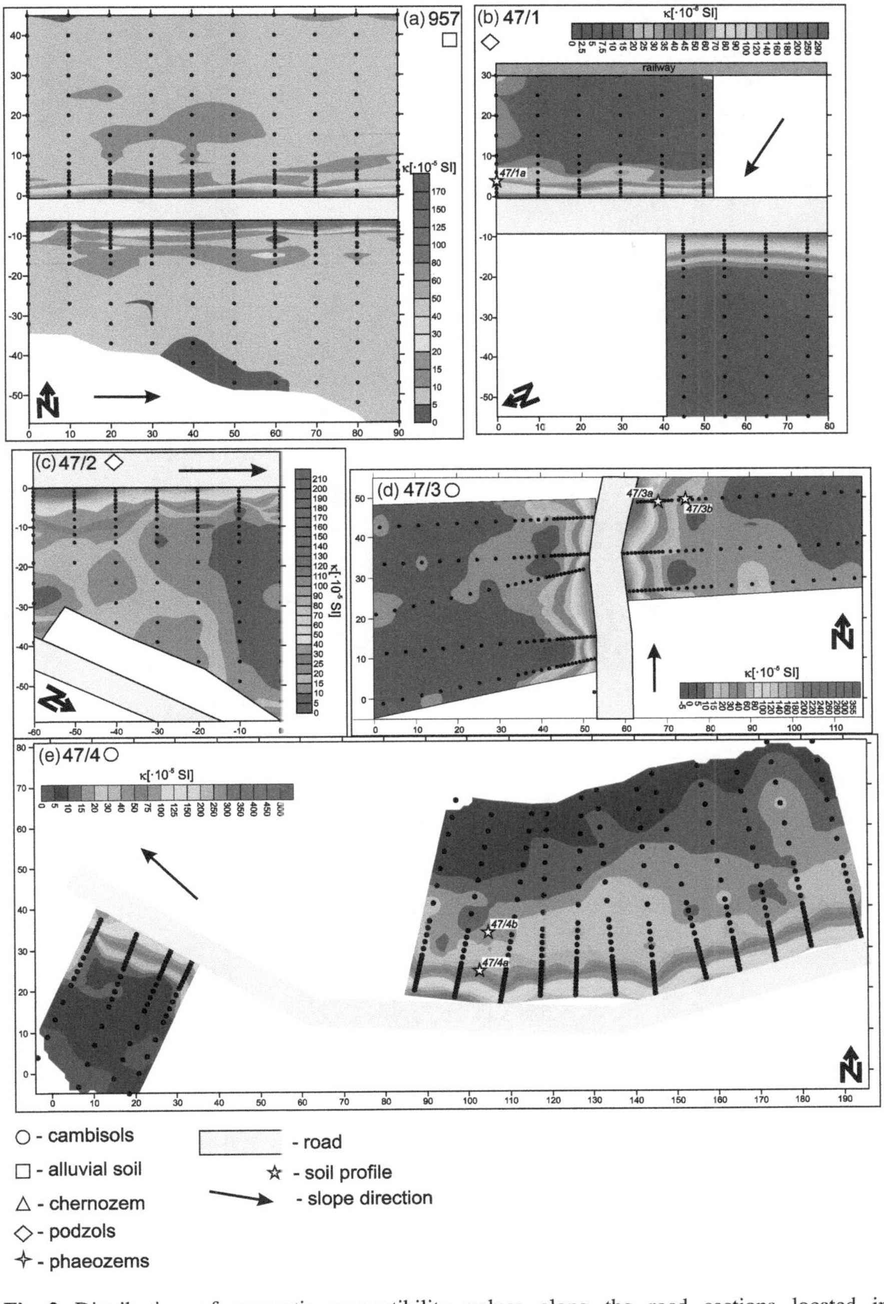

Fig. 2 Distribution of magnetic susceptibility values along the road sections located in Carpathians. Sites of measurements are marked by *black* points

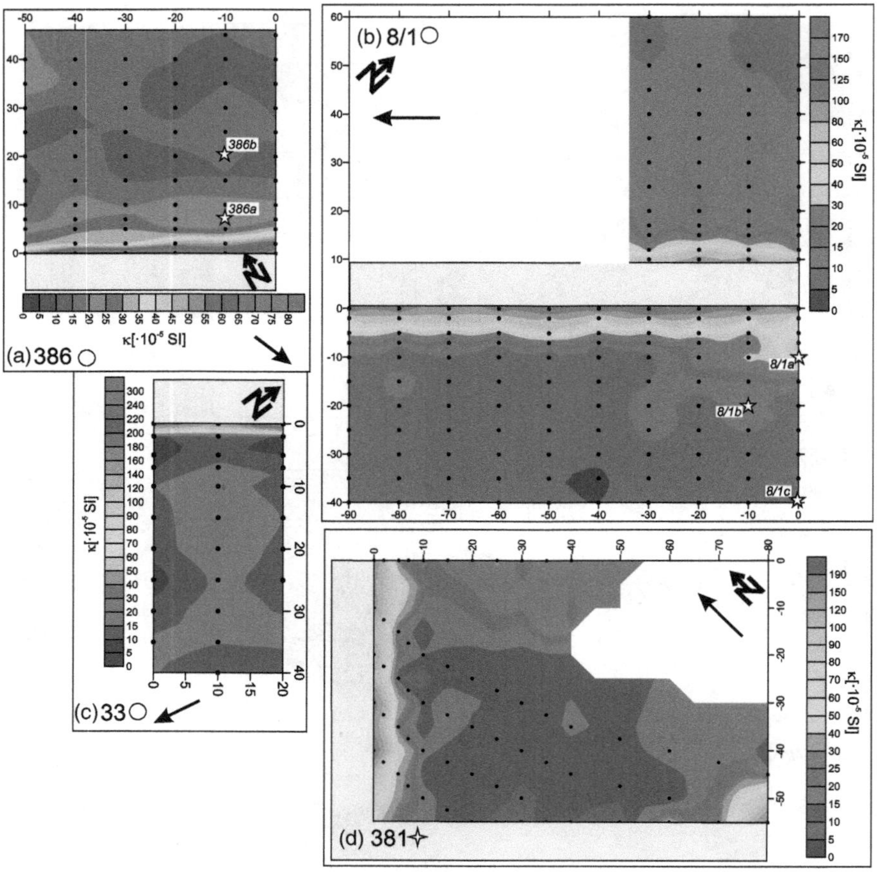

Fig. 3 Distribution of magnetic susceptibility values along the road sections in the Sudetes

776) is characterized by highest susceptibilities among all studied soils ($\kappa_{back} \approx 40 \times 10^{-5}$ SI). The values of κ_{back} of other soils oscillate around 10–15 $\times$ 10^{-5} SI. However, it is impossible to establish one value of κ_{back} for individual soil type since it is strongly dependent on soil basement.

Concise descriptions of magnetic susceptibility distribution for all road sections are summarized in Table 1. The summary shows that in most cases the width of distinctly increased κ bandpass equals ca. 10 m. As expected, the widest zone of increased κ is situated by A4 motorway and the narrowest one surround subordinate roads e.g. 776 ($\sim$1 m) and 957 ($\sim$3 m).

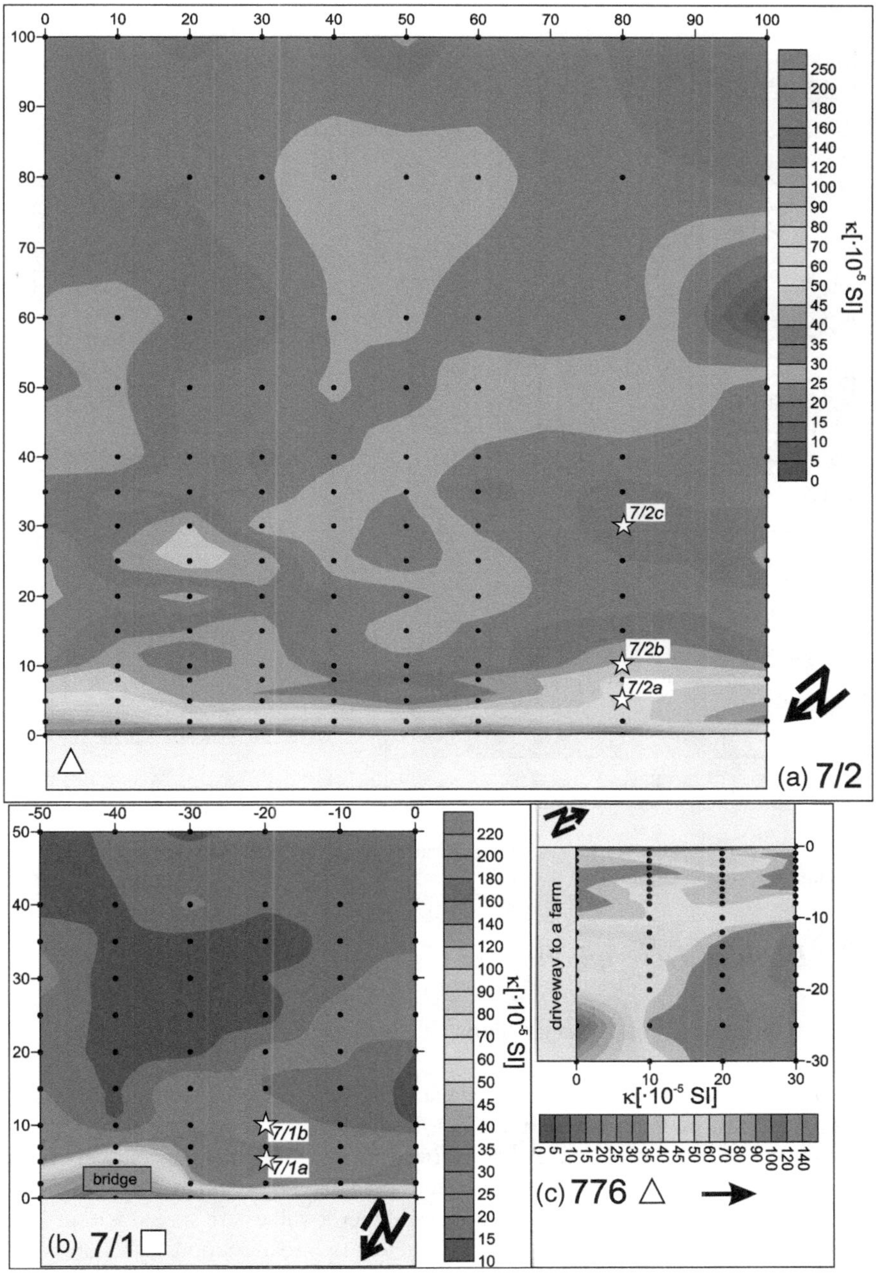

Fig. 4 Distribution of magnetic susceptibility values along the road sections in Małopolska Upland

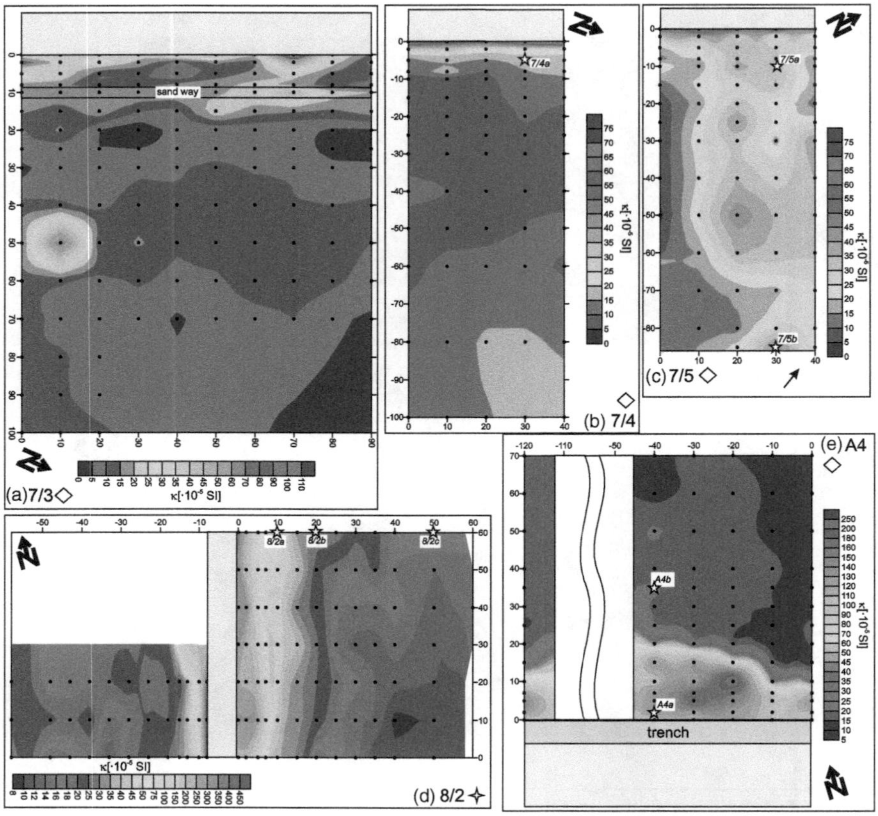

Fig. 5 Distribution of magnetic susceptibility values along the road sections located at Polish lowlands

3.2 *Detailed Description*

3.2.1 Mountain Area—Carpathians

Four road sections were chosen at the main road no. 47 connecting Cracow with Zakopane. Two of them (47/1 and 47/2), farthest to the south, are placed among podzolic soils, which usually are characterized by quite low magnetic susceptibility. The maximum value of noncontaminated podzols is established on the level $\kappa_{back} = 15 \times 10^{-5}$ SI. At both road sections higher κ values are spread within 10 m bandpass along the road (Fig. 2b and c). The next two road sections (47/3 and 47/4) are placed among cambisols that are characterized by similar κ values as the podzolic soil. At the 47/3 section (Fig. 2d) the values of κ higher than 15×10^{-5} SI are extended up to 15 m from the road edge. This section is quite flat and the shape of magnetic susceptibility anomaly is symmetric with respect to the road axis.

Table 1 Magnetic susceptibility, with of anomaly band, road slope and soil type characteristic for particular road sections

	Name place	Road type	Road slope (%)	Soil type	κ_{max} ($\times 10^{-5}$ SI)	κ_{back} ($\times 10^{-5}$ SI)	$\Delta\kappa = \kappa_{max} - \kappa_{back}$ ($\times 10^{-5}$ SI)	Width of anomaly band (m)	Pollution
Carpathians	47/4	Primary 2-lanes	6.7	Cambisols	496	~20	476	~10	Strong
	47/3	Primary 2-lanes	1.0	Cambisols	320	~20	300	~15	Strong
	47/1	Primary 2-lanes	1.7	Podzols	282	~15	267	~10	Strong
	47/2	Primary 2-lanes	1.3	Podzols	203	~20	183	~10	Moderate
	957	primary 2-lanes	2.0	Alluvial soils	166	~10	156	~3	Moderate
Sudetes	33	Primary 2-lanes	2.6	Cambisols	290	~15	275	~2	Strong
	381	Primary 2-lanes, crossroads	2.2	Phaeozems	186	~15	171	~8	Moderate
	8/1	Primary 3-lanes	3.8	Cambisols	165	~15	150	~10	Moderate
	386	Primary 2-lanes	1.8	Cambisols	69	~20	49	~10	Weak
Malopol. upl.	7/2	Primary 2-lanes	1.1	Chernozem	250	~45	205	~10	Strong
	7/1	Primary 2-lanes	1.1	Alluvial soils	213	~25	188	~10	Moderate
	776	Primary 2-lanes	4.2	Chernozem	138	~40	98	~1	Weak
Polish lowl.	7/4	Primary 2-lanes	0.8	Podzols	75	~20	55	~3	Weak
	7/5	Primary 2-lanes	1.8	Podzols	71	~45	26	~3	Weak
	7/3	Trunk r., dual carriageway	0.8	Podzols	109	~15	94	~15	Weak
	8/2	Primary, dual carriageway	0.3	Phaeozems	450	~30	420	~10	Strong
	A4	Motorway, dual carriageway	2.6	Podzols	250	~20	230	~25	Strong

The situation is different in the case of 47/4 road section (Fig. 2e), the steepest among all studied locations. It traverses a slope of a hill. That is why an asymmetric distribution of increased κ was detected there. On the northern side, where the slope descends gently, the value of $\kappa_{back} = 15 \times 10^{-5}$ SI appears farther than 20 m from the road edge while on the southern side, where the slope rises above the road level, the same value of κ_{back} is present at 10 m of distance.

The fifth road section was located at the subordinate road no. 957 between Zakopane and Jabłonki. It was constructed on alluvial deposits. Increased values of κ are present at the road earthworks only and extend no farther than 3 m from the road edge.

3.2.2 Mountain Areas—Sudetes (Kłodzko Basin)

We selected here two sections of subordinate roads. The first is located on the road no. 381, which is surrounded by the phaeozems soil. The section is located close to a crossing of Kudowa-Bardo road and Kłodzko-Wałbrzych road. On the east side, this road section borders on a meadow. The studied area was limited by the hop field. It is slightly lowered and traces of flooding were visible. Very low values of κ were detected. It is likely that flooding and marshy soil environment are not conducive to maintaining magnetic particles containing iron in higher oxidation states. A bandpass of increased κ is very narrow there. It does not reach over 5–8 m (Fig. 3).

All other roads in the Sudetic group are located at cambisols. Around the section of secondary road no. 386 (Fig. 3a) the cambisol is ca. 30 cm thick with red clay at its basement. The value of κ_{back} in the studied area was established at the level of $20–10^{-5}$ SI. The higher values of κ were detected up to 10 m from the road edge.

Two other Sudetic sections are located at the main roads no. 33 (Fig. 3c) and no. 8 (Fig. 3b). At the section of road no. 33, evidently increased values of κ ($>60 \times 10^{-5}$ SI) were observed in the immediate vicinity of the road only, i.e., up to 2 m from the edge of the asphalt. Further away, the κ values within the range of 10–20 $\times 10^{-5}$ SI were dominant. Such values are typical for un-contaminated cambisols. It is possible that the anomaly band is fairly narrow because there is a trench and row of trees separating a road from the measured field. As it is known, trees accumulate some road pollution (Matzka and Maher 1999). At section of road no. 8 located between Kłodzko and Bardo, a bandpass of increased κ ($>15 \times 10^{-5}$ SI) is distinctly wider to the south-east from the road, i.e., with the dominant wind direction. A width of anomalous bandpass equals here 10 m and in some places reaches up to 15 m. At the opposite side of this road, the increased κ values extend only to 5 m distance from the road edge.

3.2.3 Upland Roads

Two locations at main road no. 7, connecting Cracow and Zakopane (sections 7/1 and 7/2) and part of a subordinate road no. 776 were the subject of our study in the upland group. The section 7/1 (Fig. 4b) was located among alluvial soils developed on the loess deposits. Such a type of basement can provide relatively higher values of $\kappa_{back} \approx 25 \times 10^{-5}$ SI. They were detected no closer than 10 m from the road. Significantly increased values of $\kappa > 50 \times 10^{-5}$ SI were noted at first 2–5 m of the road surroundings only.

The next two road sections were located on chernozems. At the section no. 7/2 (Fig. 4a), data were collected on the north-west side of the road. The chernozems were developed on the loess. Pure, natural chernozems are characterized by relatively high values of magnetic susceptibility. The values of $\kappa \approx 45 \times 10^{-5}$ SI were noted in many places in the area. A source of such high values is not connected with anthropogenic pollution at all but with natural magnetic enhancement and they are typical for both young and paleochernozems. Values of κ higher that 45×10^{-5} SI were measured at a distance not greater than 10 m from the road edge.

The situation is different in the surroundings of a subordinate road no. 776 (Fig. 4c). There is no magnetic susceptibility enrichment in the background of 1 m bandpass caused by road substructure, but some high values of κ (up to 80×10^{-5} SI) were detected in many other places of the studied area.

3.2.4 Lowland Roads

Three of five lowland sections are located at the main road no. 7 (sections: 7/3, 7/4 and 7/5). The other two sections, 8/2 and A4, were chosen in the vicinity of Wroclaw. With the exception of the 8/2 road section, which is located at phaeozems, other sections are surrounded by podzols.

The 7/3 section (Fig. 5a) was chosen at a newly built trunk road connecting Warsaw with Radom. Data were collected on the east side of the road. The anomalies of magnetic susceptibility are irregular. Their shape is probably affected by the heterogeneity of a ground (fallows, plowed field, sandy road). There is no bandpass of increased κ caused by the road substructure, as it is visible on previous maps, since this time the substructure is built of sand, not of gravel. These results will form a background for further measurements in the future.

Data at the podzolic 7/4 section (Fig. 5b) were collected on the east side of the road. The values of $\kappa > 20 \times 10^{-5}$ SI are detected at a distance up to 3 m from the road edge only. Conceivably anomalous values are present a bit farther, up to 20 m at most, where $\kappa > 10 \times 10^{-5}$ SI. Further away from the road, susceptibility decreases but at a ca. 60 m distance it grows again. Such a pattern seems to reflect the presence of glacier moraine in the basement of the soil.

There are moraines under podzols at the section 7/5 (Fig. 5c). The values of κ up to 45×10^{-5} SI are found in many places of the studied area. These increases are caused most probably by the iron oxides released from the moraine material. The higher values of magnetic susceptibility were noted there within a narrow (3 m) bandpass at the road earthworks only.

The section A4 (Fig. 5e), i.e., the first of two sections in the vicinity of Wroclaw, is the only section of motorway. Structure of the motorway (concrete ditch) made it impossible to survey and sample in a direct neighborhood of the road. The values of κ at 35 m of distance are stable and do not grow above $\kappa_{back} = 20 \times 10^{-5}$ SI (with one exception at the southern profile only). A bandpass of a very high κ (up to 280×10^{-5} SI) extends up to 25 m from the road edge. The border of increased κ coincides with the border of gravel foundation of the motorway.

Value of κ_{max} results mainly from an increased content of anthropogenic particles. Therefore, it can achieve a similar, high level independently of the original soil material. If A4 and 7/2 had a similar value of κ_{max} (Figs. 4a and 5d), but after subtraction A4 has higher value of $\Delta\kappa$ than 7/2 (Table 1), this may indicate that at A4 there is higher accumulation of anthropogenic particles than at 7/2, but both places are strongly polluted.

The 8/2 section (Fig. 5d) is localized at the main road no. 8, among phaeozems. The values of $\kappa > 30 \times 10^{-5}$ SI are considered as anomalous in this case. A width of anomalous κ reaches 10 m on the west side of the road, and it is twice wider on the east. Since the area is flat, the main factor causing such a difference seems to be the prevailing western wind direction.

3.3 *Vertical Profiles*

The main aim of soil profiling was to check a vertical extension of pollution. Table 2 presents features of each soil profile. There is no significant relation between depth of migration of pollution and type of soil. Maximum depth of anomalous χ (higher than susceptibility of background χ_{back}) varies between 10 and 50 cm. The rule is that farther away from the road the depth of anomalous χ is smaller (see e.g. profile 7/2—Fig. 6f, profile 386—Fig. 6d and profile 47/4—Fig. 6c). Results of laboratory measurements of magnetic susceptibility depend clearly on the position of the chernozem profiles 7/2a and 7/2b in relation to κ anomaly. The values of χ measured on samples taken from the farthest soil profile (7/2c) display only minor changes. The presence of fine-grained most probably pedogenic magnetite is indicated by high values of χ_{FD}. Data from the rest of profiles (7/2a and 7/2b) are evidently different. Increased values of χ at the top of these sections are accompanied by low values of χ_{FD}. This fact indicates that the magnetic susceptibility is here carried out by larger grains. They cannot be linked with natural pedogenic processes.

All samples from the profile 386a located near the road display increased values of magnetic susceptibility and the χ_{FD} parameter. It is possible that magnetic mineralogy was affected here also by the road construction material.

In the area of the section no. 8/1, three soil profiles were studied (Fig. 6e). One profile (8/1a) was located inside the anomaly of magnetic susceptibility. The maximum value of magnetic susceptibility is noted in the sample from the surface. Both sections located behind the anomaly of magnetic susceptibility show this maximum at depths of 25 and 10 cm. These highest values are here significantly lower than noted in the profile 8/1a.

Table 2 Magnetic susceptibility and vertical extension of its anomaly in particular soil profiles

Profile	Soil type	χ_{max} ($\times 10^{-8}$ m^3 kg^{-1})	Depth of χ_{max} (cm)	χ_{back} ($\times 10^{-8}$ m^3 kg^{-1})	Vertical extension (cm)	In- or outside of anomaly?
47/1a	Podzols	36.50	10	15	10	In
47/3b	Cambisols	16.23	5	15	15	Out
47/4a	Cambisols	87.00	10	15	15	In/out
47/4b	Cambisols	18.02	10	15	10	Out
386a	Cambisols	56.67	30	20	30	Out
386b	Cambisols	30.33	15	20	15	Out
8/1a	Cambisols	37.17	0	10	20	In
8/1b	Cambisols	20.17	25	10	25	Out
8/1c	Cambisols	22.67	10	10	20	Out
7/2a	Chernozem	70.83	10	35	20	In
7/2b	Chernozem	51.33	10	35	20	In/out
7/2c	Chernozem	47.83	5	35	5	Out
7/1a	Alluvial soil	36.33	10	20	15	Out
7/1b	Alluvial soil	29.33	10	20	10	Out
7/4a	Podzols	17.32	20	10	30	In/out
7/5a	Podzols	39.67	10	15?	30	In/out
7/5b	Podzols	40.33	10	15?	20	Out
8/2a	Phaeozems	81.83	20	15	50	In
8/2b	Phaeozems	79.00	30	15	30	In/out
8/2c	Phaeozems	30.83	10	15	20	Out
A4b	Podzols	32.67	5	15	35	Out

Diversity of magnetic susceptibility patterns in the profiles studied at the section no. 8/2 is mainly due to different thickness of soil accumulation horizon in each profile and different kind of the bedrocks. The increased values of magnetic susceptibility cover the whole humus horizon (50 cm) cropping out in trench no. 8/2a that was dug c.a. 10 m from the road edge.

3.4 *Magnetic Mineralogy*

The IRM curves indicate (Fig. 7) that low-coercivity magnetic minerals are present inside the magnetic susceptibility anomalies. Especially, the samples from the surface and near-surface display the lowest coercivities. This is clearly visible for example in the case of the section no. 386 (Fig. 7c and d). Completely different situation is observed in the profile no. A4a (Fig. 7i) where all samples contain mainly low-coercivity minerals.

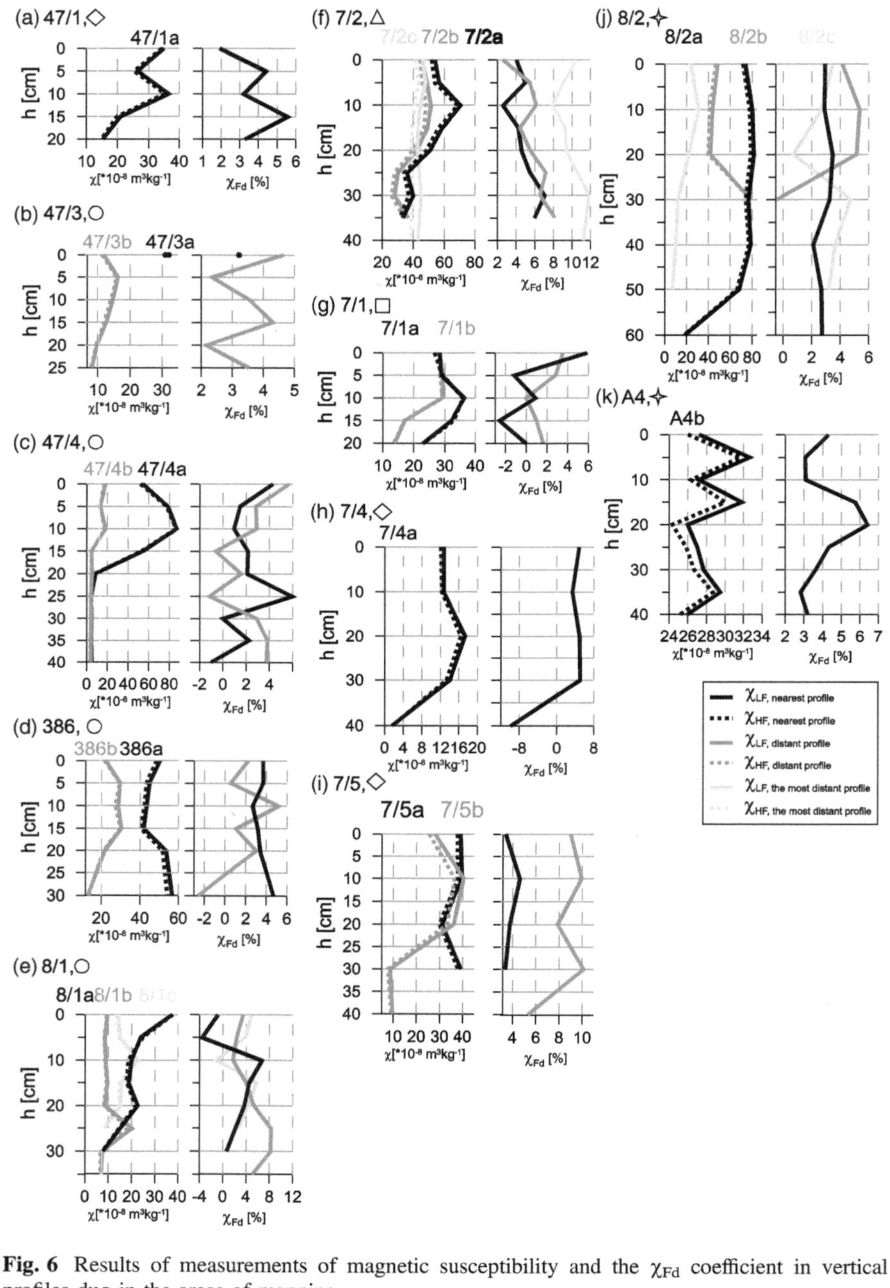

Fig. 6 Results of measurements of magnetic susceptibility and the χ_{Fd} coefficient in vertical profiles dug in the areas of mapping

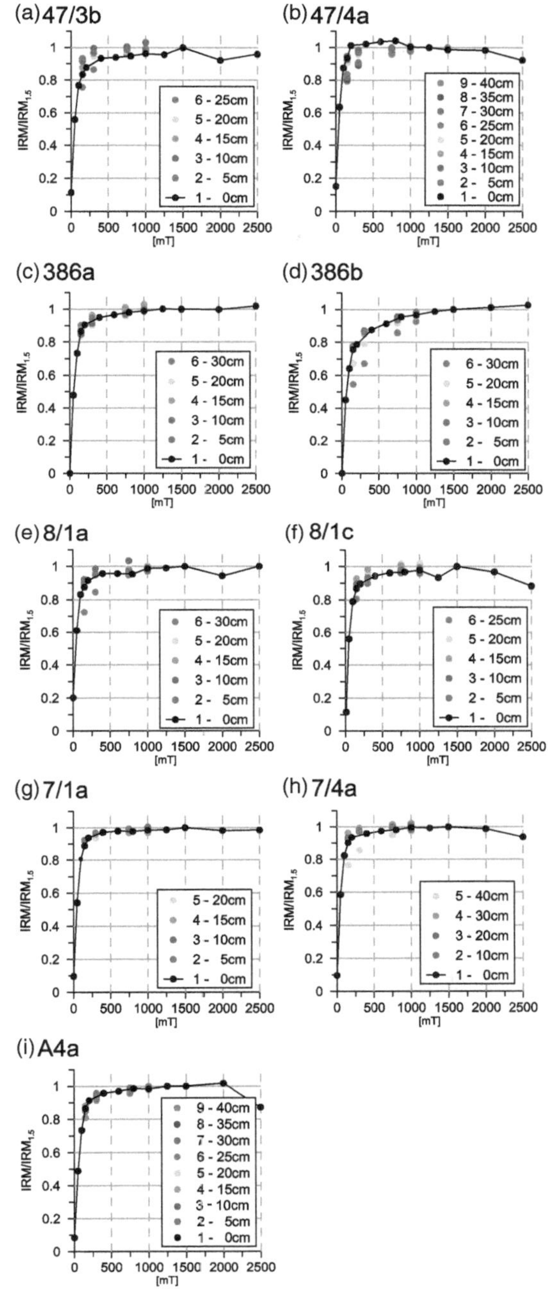

Fig. 7 The isothermal remanent magnetization graphs prepared for selected samples taken from different depths of studied soil profiles

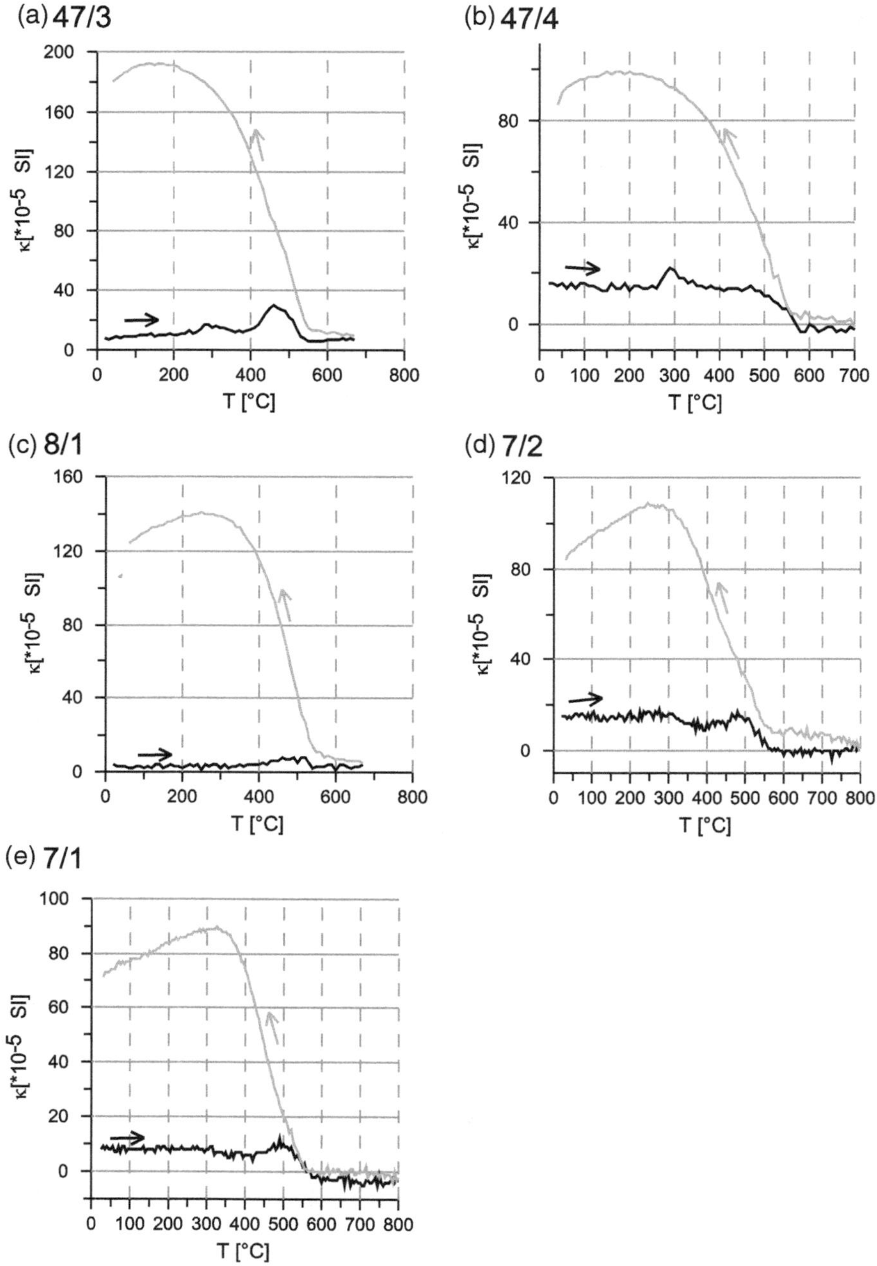

Fig. 8 Normalized susceptibility versus temperature heating and cooling curves prepared for selected soil samples

Thermomagnetic curves indicate that magnetite is the main magnetic mineral in all studied samples. A substantial decrease of magnetic susceptibility is visible around temperature of about 580 °C (Fig. 8).

In some samples, a certain addition of maghemite is also possible. It is indicated by a decrease of magnetic susceptibility at temperatures higher than 300 and lower than 450 °C (Fig. 8a, d, and e) due to oxidation of this mineral to less susceptible hematite. A significant increase of magnetic susceptibility in some samples at temperature of c.a. 480 °C (Fig. 8a) can be linked with the Hopkinson peak and the new magnetite grains formation during heating from Fe-rich clay minerals.

3.5 SEM Images

The magnetic particles observed on SEM images can be subdivided into two groups. The first group contains magnetic spherules of different kind (Fig. 9). The smallest of them have a diameter of c.a. 40 μm and the largest even 160 μm. Part of magnetic spherules are filled with material containing Si, Ca, Al and Mg. The other ones are empty inside. Traces after gas bubbles are visible inside the internal structure of some spherules. The bubbles were most probably formed during combustion. The second group of magnetic particles contains irregular elements

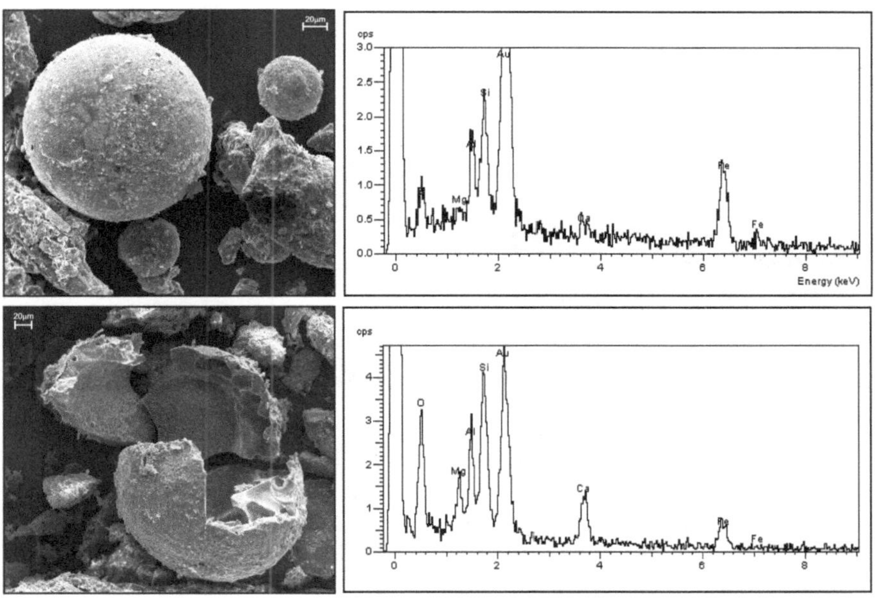

Fig. 9 SEM image of ferric spherules extracted from polluted soils and distribution of elements on their surface and inside them (*lower diagram*). The content of Au is not coming from spherules. This element was used in preparatory process

with sharp or oval margins (Fig. 10). Flat particles contain iron and aluminum. In some cases, the titanium was also found. Si and Al usually occur in material filling the particles. The irregular particles most probably are coming from the dust propagated by cars. The anthropogenic nature of spherules is supported by the fact that they are absent in the profiles located behind magnetic susceptibility anomalies far away from the roads.

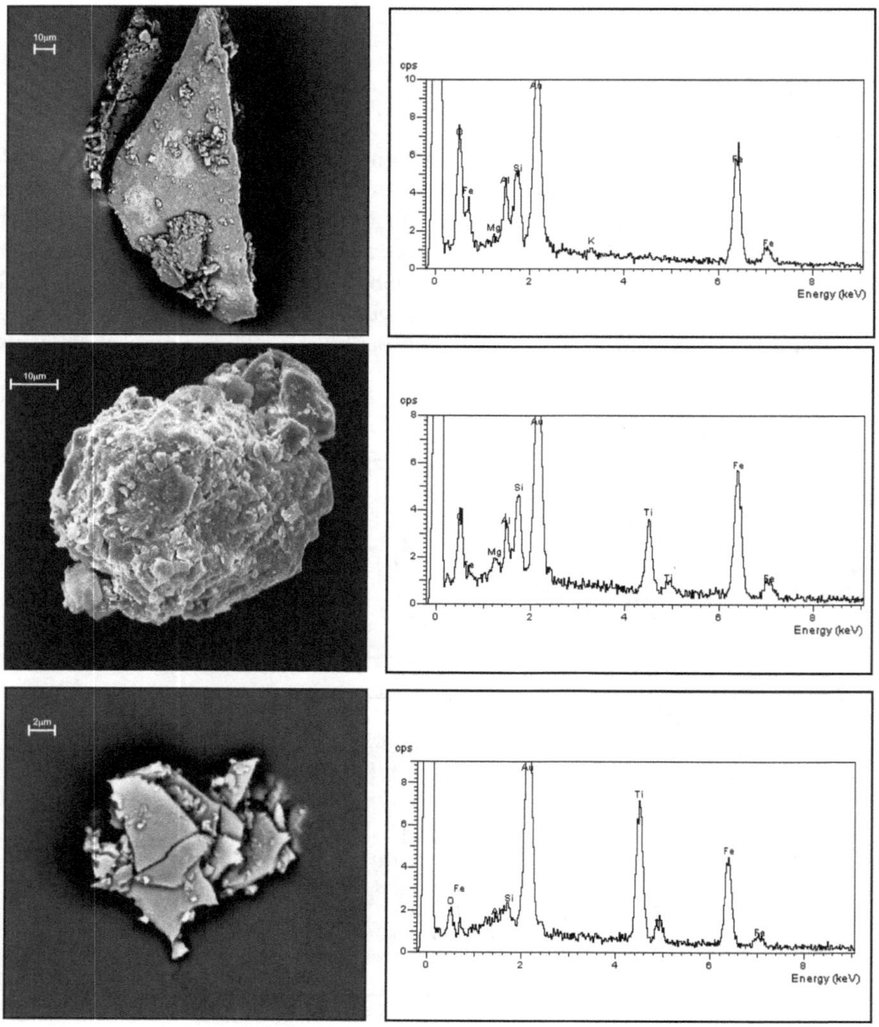

Fig. 10 SEM image of irregular containing iron particles extracted from polluted soils and distribution of elements on their surface

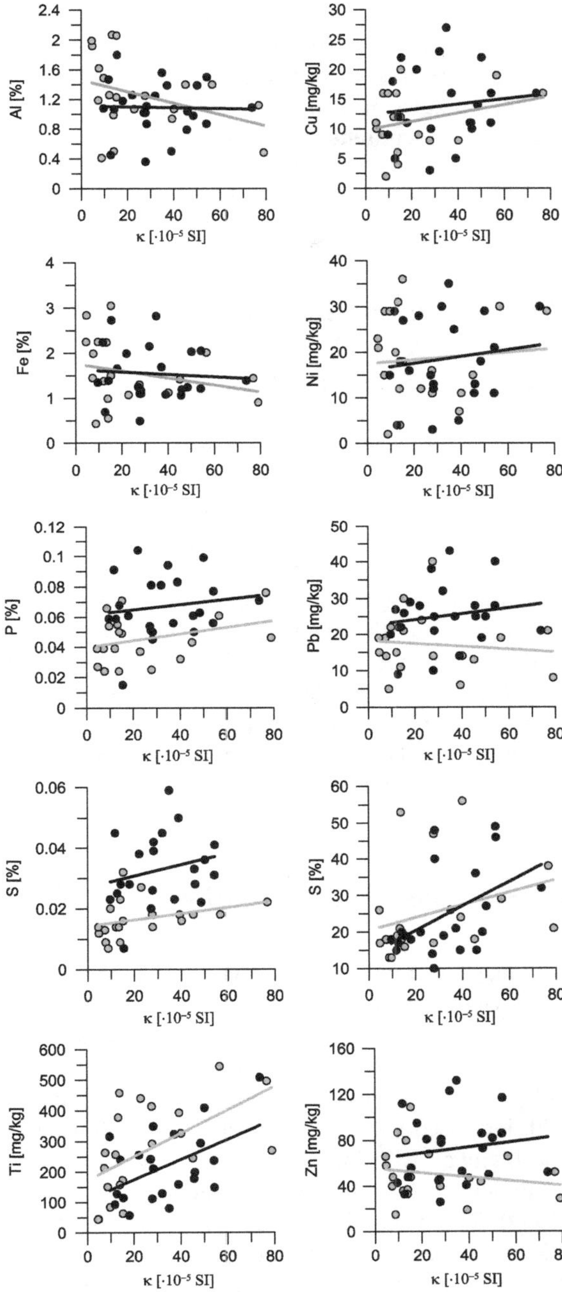

Fig. 11 Magnetic susceptibility of polluted soils versus content of selected elements. *Black* points represent samples taken in the most polluted soils displaying highest values of magnetic susceptibility. *Grey* colored points represent soil samples from more distant parts of magnetic susceptibility anomalies

In the sample taken from the surface of the profile A4b, the SEM study indicates titanomagnetite as the main magnetic mineral. It forms irregular grains. The crystal walls are preserved very rarely. Significantly larger magnetic spherules contain here small admixture of titanium. It can be concluded that close to the A4 highway the magnetic susceptibility anomalies are caused by magnetic particles not only from anthropogenic source (spherules) but also from natural gravels (titanomagnetites) forming the ground of the road.

3.6 Magnetic Susceptibility and Geochemistry

Results of geochemical analyses of samples taken from the surface and at a depth of 30 cm were correlated with magnetic susceptibility of the same samples (Fig. 11). The concentration of many elements depends on the depth of sampling. The samples from the lower horizon of sampling are richest in titanium. It is built mainly in natural titanomagnetite grains. In both compared horizons, the total content of titanium evidently correlates in a positive way with the magnetic susceptibility. Soil forming processes reduce content of detrital components in accumulation horizon. Apart from titanium the rest of elements showing positive correlation with the magnetic susceptibility (e.g. sulphur and phosphorus) display more distinct one in the sample set taken on the surface. The upper horizon of sampling is more abundant in zinc and lead than the lower one. The content of these elements is positively correlated with the magnetic susceptibility values in the samples from the top of soil only. The occurrence of zinc and lead can be linked therefore with anthropogenic delivery.

4 Summary and Discussion

The analysis of magnetic susceptibility distribution at the farmlands which lie in the close vicinity of roads shows that there is a bandpass of increased susceptibility along each studied road section. The bandpasses have variable width. In the case of subordinate roads with smaller traffic, the increased values of magnetic susceptibility were noted in hard shoulder i.e. up to 3 m from the road edge. A wider belt with increased values of magnetic susceptibility is observed along the main roads and highway with intense traffic. It reaches somewhere 40 m, but usually covers 15 m only. These values are substantially larger than those noted by Hoffmann et al. (1999). The belts with anomalous values of magnetic susceptibility were defined thanks to measurements conducted also far away from the roads. In the case of cambisols, natural values of magnetic susceptibility do not exceed 20×10^{-5} SI. The higher values of this parameter (up to 45×10^{-5} SI) were noted in natural environment of chernozems and soils developed on moraines.

Increased traffic-inducted values of magnetic susceptibility were observed up to 30 cm of depth. Below this depth the magnetic susceptibility signal is typical for the soil bedrocks in the whole area of the particular road section. Magnetite is the main magnetic susceptibility carrier. Metallic iron and titanomagnetite were also identified. This supports earlier observations (Kim et al. 2007; Bućko et al. 2011; Jordanova et al. 2014). The iron-rich spherules were found frequently in the profiles located inside the magnetic susceptibility anomalies. Because of this, their origin cannot be constrained to industrial activities and household heating systems, as postulated by Bućko et al. (2011).

Fragments of crystal walls have been preserved in some grains containing titanium and iron. Many of particles with these elements occur in the shape of elongated chips. The first population contains most probably a detrital titanomagnetite that occurs frequently in gravel and sand used for construction of the roads. This material undergoes erosion and transportation towards the farmlands. Irregular forms containing titanium and iron and ferric spherules are coming from the traffic. In the nearest surroundings of roads, high values of magnetic susceptibility are linked with both sources i.e. the traffic and the ground used for construction of roads. Far away from the roads, still anomalous but lower values of magnetic susceptibility are caused by the traffic only. This observation partly does not support the conclusion of Jordanova et al. (2014) who describe traffic source of increased susceptibility only.

The susceptibility carriers from the samples taken in the anomalous areas are associated with highest content of zinc, lead, sulphur, and to a lesser degree—nickel. Increase of magnetic susceptibility is accompanied by increase of content of these elements in the topsoil. Such a relationship was also observed by Bućko et al. (2011). Their raised concentration in farmlands can be dangerous for human and animal health. Because of this, a quick and cheap detection of anomalous areas is of great importance. Our data show that the zone up to 15 m, and somewhere even up to 40 m distant from the roads should be excluded from agriculture. Unfortunately, meadows and farmlands are usually very close to the road edge.

5 Conclusions

- A belt with increased values of magnetic susceptibility observed along the main roads and highway with intense traffic in Poland reaches somewhere 40 m but usually covers 15 m of the neighboring areas.
- Increased traffic-inducted values of magnetic susceptibility were observed up to 30 cm of depth of the soil profile.
- In the closest vicinity of roads high values of magnetic susceptibility are caused by the traffic and material used for construction of roads.

- Increase of magnetic susceptibility is accompanied by increase of content of zinc, lead, sulphur, and to a lesser degree—nickel in the topsoils surrounding the roads.
- The zone up to 15 m, and somewhere even up to 40 m away from the road edges should be excluded from cultivation.

References

Beckwith PR, Ellis JB, Revitt DM, Oldfield F (1986) Heavy metals and magnetic relationships for urban source sediments. Phys. Earth Planet Inter. 42:67–75

Beckwith PR, Ellis JB, Revitt DM (1990) Applications of magnetic measurements to sediment tracing in urban highway environments. Sci. Total Environ. 93:449–463

Bućko MS, Magiera T, Johanson B, Petrovsky E, Pesonen LJ (2011) Identification of magnetic particles in road dust accumulated on roadside snow using magnetic, geochemical and micromorphological analyses. Environ. Polut. 159:1266–1276

Goddu SR, Appel E, Jordanova D, Wehland F (2004) Magnetic properties of road dust from Visakhapatnam (India)—relationship to industrial pollution and road traffic. Phys. Chem. Earth 29:985–995

Hoffmann V, Knab M, Appel E (1999) Magnetic susceptibility mapping of roadside pollution. J. Geochem. Explor. 66:313–326

Jordanova D, Jordanova N, Petrov P (2014) magnetic susceptibility of road deposited sediments at a national scale—relation to population size and urban pollution. Environ. Pollut. 189:239–251

Kim W, Doh SJ, Park YH, Yun ST (2007) Two-year magnetic monitoring in conjunction with geochemical and electron microprobe data of roadside dust in Seoul. Korea. Atm. Environ. 41:7627–7641

Kletetschka G, Zila V, Wasilewski PJ (2003) Magnetic anomalies on the tree trunks. Studia Geophys. Geodaet. 47:371–379

Lu S, Bai S, Cai J, Xu C (2005) Magnetic properties and heavy metal contents of automobile emission particulates. J. Zhejiang Univ. Sci. 6B(8):731–735

Matzka J, Maher BA (1999) Magnetic biomonitoring of roadside tree leaves: identification of spatial and temporal variations in vehicle-derived particles. Atm. Environ. 33:4565–4569

Magnetic Study of Sediments from the Vistula River in Warsaw—Preliminary Results

Iga Szczepaniak-Wnuk and Beata Górka-Kostrubiec

Abstract The aim of the study was to identify magnetic particles in the sediments of Warsaw part of the Vistula river (Poland) and the impact of anthropogenic factors on the magnetic susceptibility. One of the most important parts of this study was to determine the procedure for sediment sampling and preparation of material for magnetic measurements. The paper presents results of the preliminary study on the test collection of sediment samples from the Vistula river bank within Warsaw. The main magnetic study was carried out on the collection of sediments taken from six areas with different impact of anthropogenic sources of pollution using the procedure developed for test collection. For samples of sediments, several magnetic parameters, such as magnetic susceptibility χ, temperature dependent magnetic susceptibility, and hysteresis loop parameters, were determined. The study showed that the magnetic susceptibility of sediments reflects the anthropogenic impact. In the city center, the magnetic susceptibility reached maximum values which reflect the highest anthropogenic factor. The lowest magnetic susceptibility values were recorded in sediments taken from out of the city center areas which shows relatively low impact of anthropogenic factor. The fine-grained fractions of sediments contain most of the magnetic particles in which multidomain magnetite is the main magnetic phase. It was found that the spherical-shaped particles represent a significant part of magnetic fraction of sediment. The results bring us to the conclusion that the application of magnetic methods for study of river sediments is very important for understanding and tracing the anthropogenic input in the river sediments.

Keywords Magnetic susceptibility · Magnetic method · River sediments · Spherules · Vistula river

I. Szczepaniak-Wnuk (✉) · B. Górka-Kostrubiec
Institute of Geophysics, Polish Academy of Sciences, ks. Janusza 64,
01-452 Warsaw, Poland
e-mail: igasz@igf.edu.pl

M. Jeleńska et al. (eds.), *Magnetometry in Environmental Sciences*,
GeoPlanet: Earth and Planetary Sciences, DOI 10.1007/978-3-319-60213-4_2

1 Introduction

The magnetic method—magnetometry—is used to identify the magnetic particles presented in different pollutants. The method bases on the measurement of magnetic parameters of pollution containing the magnetic iron-oxides and iron sulfides. The magnetic parameters very often show correlations with concentration of heavy metals and other toxic trace elements. The magnetic method is a very fast, sensitive, and inexpensive tool to identify the pollution levels of soil (Jelenska et al. 2008; Bućko et al. 2010), atmospheric air by measuring particulate matter (PM) (Sagnotti et al. 2006; Castaneda-Miranda et al. 2014) and indoor air by measuring household dust (Górka-Kostrubiec et al. 2014; Górka-Kostrubiec 2015; Szczepaniak-Wnuk and Górka-Kostrubiec 2015).

The motivation to undertake the research is the link documented by many authors between river sediments pollution levels and their magnetic parameters.

Petrovský et al. (2000) analyzed the sediments of the Vltava river (Czech Republic) in the terms of impact of local pollution sources on the magnetic susceptibility of sediments. It has been shown that measurement of low-field magnetic susceptibility can be used to map the areas exposed to various sources of pollution and to investigate the ways of pollution transfer in different ecosystems.

Chaparro et al. (2003, 2004) investigated the effect of anthropogenic factors on magnetic enhancement in sediments collected from two rivers, Del Gato and El Pescado, in Argentina. In the sediments of Del Gato river, flowing through the city center, they recorded higher magnetic susceptibility, than in the sediments of El Pescado river flowing through rural areas. The magnetic fractions of both river sediments contained mainly soft magnetic magnetite. In addition, in the sediments of El Pescado river the harder magnetic phase, probably of lithogenic origin hematite, was found. The authors obtained a very good correlation of the magnetic susceptibility with the concentration of heavy metals.

The investigation reported by Desenfant et al. (2004) was carried out in order to identify the anthropogenic origin of magnetic particles in river sediments, and to find possible correlations between magnetic parameters and heavy metals content. They established the relations between the magnetic anomalies of sediments and the presence of lead, iron, zinc and chromium elements coming from the industrial activity. The iron-rich spherical particles corresponding to particles produced during the combustion process were found around the thermal power plant. The spherules were considered as a major pollutant associated with human activity, because they were not observed in sediments from the upper part of the river, where the impact of the thermal power plant was neglected.

Jordanova et al. (2004) studied the magnetic fraction of sediments from the Bulgarian part of the Danube river. In the work, the microscopic, magnetic and micro-chemical properties of individual grains of polluted sediments were analyzed in detail. The magnetic fraction of deposits contained the spherules which varied in morphology and structure of surface. In most of the cases, the spherical particles had a complex internal structure containing variable chemical elements. Magnetic analysis revealed the magnetite as the main ferromagnetic phase in the river sediments.

Knab et al. (2006) studied the impact of anthropogenic activities on the sediments of the Vltava river (Czech Republic) sediments collected in the areas with complex geological structure. The magnetic susceptibility of the sediments strongly depended on the sources from which the material was provided. In the part of the river with a small lithogenic contribution, the magnetic enhancement was the result of the contribution of anthropogenic magnetic particles. Significant changes in magnetic susceptibility were observed in the part of the river mostly formed by bedrock. In this area, the local geological background was the dominant contribution to the magnetic enhancement.

The strong association between anhysteretic magnetic susceptibility (χ_{arm}) and heavy metals content was found in intertidal sediments of the Yangce Estuary (Zhang et al. 2007). The magnetic parameter χ_{arm} which is sensitive to single domain of magnetic particles showed significant relation with fine-grained fraction of sediments (d < 16 μm). This was explained by the role of grain-size, which controls the concentration of heavy metals. The grain-size mainly reflects the flow speed in the depositional site. Therefore, the granulometric fractions under different flow conditions can display the variable magnetic enrichment.

Li et al. (2011) studied the influence of industrial activities on the magnetic susceptibility of the fluvial deposits of the Yangce river. The newly deposited sediments had much higher magnetic susceptibility values than older sediments. The increase in the magnetic susceptibility in contemporary sediments corresponded to the fly ash content. The scanning electron microscopy (SEM) observations revealed the spherules as the main component of magnetic fraction.

The main aim of our investigation is to identify the magnetic particles in the sediment of the Warsaw part of the Vistula river (Poland). We chose the highly urbanized city with intensive traffic of cars, trams and buses on both sides of the river and with two all year operating electro-power plants. We wanted to check whether the type of magnetic particles and the magnetic susceptibility values reflect the anthropogenic sources of pollution.

At first, we wanted to determine and develop the procedure for sediment sampling and preparation of material for magnetic measurements. For this purpose, we carried out the preliminary study on the test collection of sediment samples. Further magnetic study and identification of magnetic particles were carried out according to the procedure developed for test collection of samples.

2 Methods and Measurements

2.1 Study Area

The research was conducted on the Vistula's sediments collected in Warsaw area, between July and November 2015. The sediment samples were taken from six areas located on the right or left bank of the river (Fig. 1). Area 1 is located about 16 km

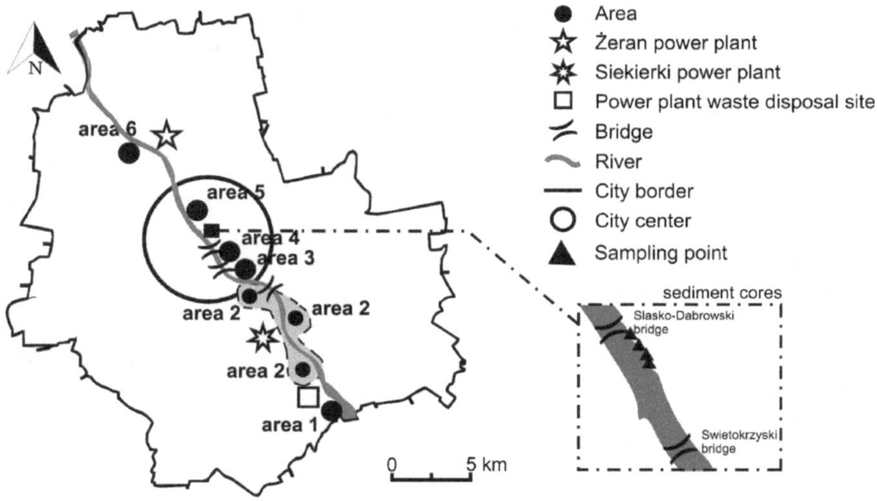

Fig. 1 The map of Warsaw with the sampling areas and the sediment cores taken between area 4 and area 5 for test study

from the city center within the Wyspy Zawadowskie preserve where the river flows into the city. The area is covered by nonregulated and naturally formed islands, sandbars and sandbanks. Area 2 is about 10 km from the city center in the vicinity of the Siekierki power plant and the power plant waste disposal site. Areas 3, 4 and 5 are in the city center along the river. Area 3 is on the left river bank between two bridges: Siekierkowski and Lazienkowski. Area 4 is between Lazienkowski and Poniatowski bridges. Area 5 is on the right bank of the river between Slasko-Dabrowski and Gdanski bridge. In the city center between two banks there is heavy traffic of buses, trams, trains and cars. Area 6 is about 7 km away from the city center (according to the course of the river) in the vicinity of the second largest power plant in Warsaw: Zeran power plant.

2.2 *Sample Collection*

The test study has been conducted for the sediment samples collected from the place located between area 4 and area 5 in the city center. There were collected 20 sediment cores using a geological core probe and, from the same places, pairs of 20 other samples of mixed material: from a depth of 0–20 cm (using a geological probe with a window) and from the surface.

The cores were taken to a depth of 20 cm, and put into closed plastic pipes ($\phi = 2\,\text{cm}$) in order to maintain the sequence of sediments. Mixed material was taken from a depth of 0–20 cm, and from the surface of sediments (area of about

0.5 m^2). All the sediment samples were strongly damped because they were taken from the places situated directly on the bank of the river.

In the laboratory, the cores were frozen to obtain a solid material which was easy to cut into small cylinders. From one core there were obtained 10 cylinders with a 2 cm height representing the following depths: 0–2 cm, 2–4 cm, 4–6 cm, 6–8 cm, 8–10 cm, 10–12 cm, 12–14 cm, 14–16 cm, 16–18 cm, 18–20 cm. The core cylinders before defrost were placed in ceramic crucibles and dried in an oven for three hours at a temperature of about 50 °C.

The samples of mixed material after the initial evaporation of water were put into ceramic crucible and also dried in an oven for 3 h at a temperature of about 50 °C. All the samples were mechanically sieved through a mesh with 1 mm openings to remove pieces of plants, stones, etc., then weighted (mass of about 1 kg) and divided into two parts. The first part was chosen for further grain size distribution analysis. The second part (m = 0.1 kg) was reserved for further study.

The division into granulometric fractions was performed using a laboratory shaker with a standard sieve set. In this study, five granulometric fractions were obtained: diameter between 1 and 0.5 mm (fraction "0.5"), between 0.5 and 0.25 mm (fraction "0.25"), between 0.25 and 0.1 mm (fraction "0.1"), between 0.1 and 0.071 mm (fraction "0.071") and less than 0.071 mm (fraction "<0.071"). The mass of each fraction was determined in order to estimate the contribution of the fraction mass to the total mass of sample. The measurement of magnetic susceptibility was performed on the arid and weighted samples which were put into plastic boxes of about 8 cm^3 volume.

The main study was performed for sediment samples collection taken from six areas along the Vistula river. For each of the six areas, several (4–6) samples were collected from the surface and from a depth of 0–20 cm in the form of mixed material. For this study, sediment cores were not collected. The manner of sampling as well as the preparation procedure (dried, sieved and division into granulometric fractions etc.) were the same as for the test collection.

2.3 Methods and Devices

For all the sediment samples, the low-field magnetic susceptibility (κ) was measured at two frequencies (f_1 = 976 Hz and f_2 = 1560 Hz) of magnetic field using Multi-Function-Kappabridge (MFK1-FA, AGICO). In environmental magnetism, the specific mass magnetic susceptibility χ is used which is defined as the magnetic susceptibility (κ) normalized on the mass of sample and expressed in $m^3\ kg^{-1}$ units.

The magnetic susceptibility is a property which describes the capability of substance to change magnetization under the influence of external magnetic field (Thompson and Oldfield 1986). The value of magnetic susceptibility provides information about magnetic mineralogy, concentration of magnetic particles and contribution of ultrafine superparamagnetic (SP) grains.

The frequency-dependence of magnetic susceptibility χ_{fd} is the parameter used to evaluate the content of ultrafine SP grains in a sample. It is defined as the difference between magnetic susceptibilities measured at low (χ_{lf}) and high (χ_{hf}) frequencies of magnetic field (Dearing el al. 1996) in relation to χ_{lf}, and expressed in %.

The changes of magnetic susceptibility κ(T) with temperature are the main magnetic feature for quick identification of magnetic mineralogy by estimation of the Curie temperature. The curves of κ(T) were measured with the Multi-Function-Kappabridge coupled with the CS-2 furnace operating in the high temperature range from 20 to 700 °C.

The shape of the hysteresis loops and their parameters are used to determine the magnetic mineralogy and domain structure of magnetic grains. For selected samples, the hysteresis loops (up to magnetic field of 1T) and the curves of subsequent DC back-field remagnetization of isothermal remanent magnetization (IRM) were measured with Vibrating Sample Magnetometer (VSM, Molspin, Great Britain). The values of saturation remanent magnetization (M_{rs}), saturation magnetization (M_s), and coercivity (H_c) were determined from the hysteresis loop and coercivity of remanence (H_{cr}) from the curve of back-field IRM.

The magnetic extracts obtained from the fine ("0.071") and the finest ("<0.071") fractions of the river sediments were analyzed by optical microscope (Nikon SMZ1500, zoom range from 0.75x to 11.25x, with digital camera).

3 Results and Discussion

3.1 *Magnetic Study of Test Collection of Sediment Samples*

Figure 2 shows an exemplary vertical distribution of magnetic susceptibility of core samples collected in the place located between area 4 and area 5. For all the samples, the magnetic susceptibility shows a similar trend; the highest values are observed for the upper layers (0–4 cm), the decrease at about 6–8 cm and, after a low increase, an almost constant value in the deeper layers (10–20 cm). It is clear that the changes of χ with depth are insignificant and fluctuate around the average value (see Fig. 2). Deviation from the average value does not exceed 15%.

Figure 3 shows an exemplary distribution of magnetic susceptibility for individual granulometric fractions collected from the surface and mixed material of sediments collected from the surface and from depth of 0–20 cm. The highest values of magnetic susceptibility have the fine "0.071" and finest "<0.071" fractions from about 15×10^{-8} m^3 kg^{-1} to about 45×10^{-8} m^3 kg^{-1} and from about 25×10^{-8} m^3 kg^{-1} to about 40×10^{-8} m^3 kg^{-1} respectively. For the coarse-grained fractions ("0.5"–"0.1") the magnetic susceptibility values are less than 5×10^{-8} m^3 kg^{-1}.

Summarizing, the changes in magnetic susceptibility with depth (Fig. 2) indicate that the highest concentrations of magnetic particles are in the surface sediment layer. For deeper layers, the changes in χ can be approximated by the average

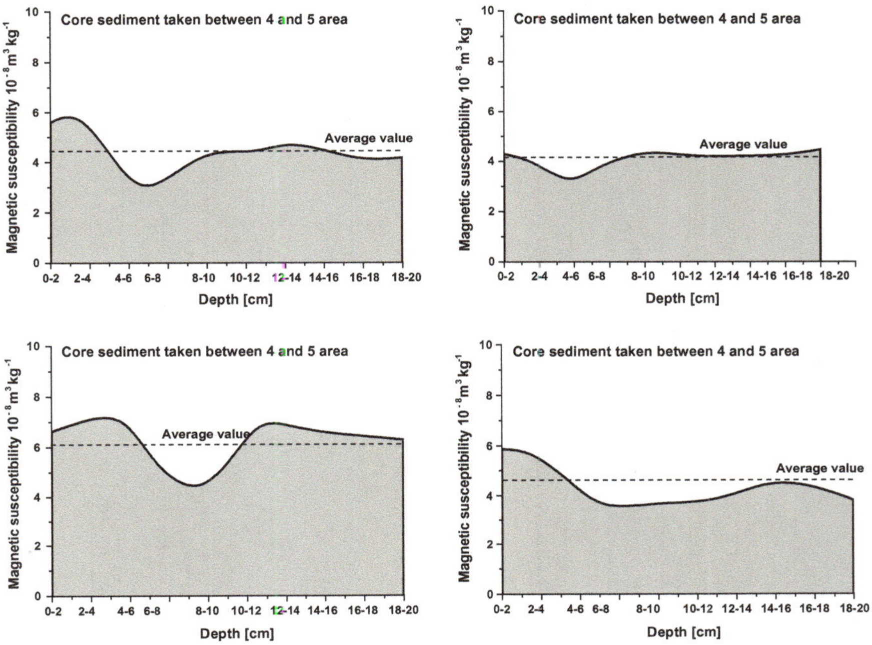

Fig. 2 Distribution of mass magnetic susceptibility (χ) of core sediments (0–20 cm) collected between area 4 and area 5

value. This result allows us to assume that for the magnetic study we should collect two samples: the first from the surface, which corresponds to the maximum enhancement in magnetic particles, and t. The second one as a mixed material from the depth of 0–20 cm, which reflects the average concentration of magnetic particles in the whole sample. We consider that the comparative analysis of magnetic properties can be successfully carried out on the fine-grained fractions of sediments which contain most of the magnetic particles.

3.2 Distribution of Magnetic Susceptibility of Sediments Collected from the Warsaw Part of Vistula River

Figure 4a shows the average values of magnetic susceptibility (χ_{av}) for fractions "0.071" and "<0.071" of surface sediments collected from each of the six areas. The lowest values are recorded in the sediments taken from the reserve zone (area 1), with a relatively low impact of anthropogenic factor. Approaching the city center, the values of χ_{av} increase, reaching a maximum at about 107×10^{-8} m^3 kg^{-1} for area 3. In the other three areas (4, 5 and 6) the values of χ_{av} fluctuate around 44×10^{-8} m^3 kg^{-1}.

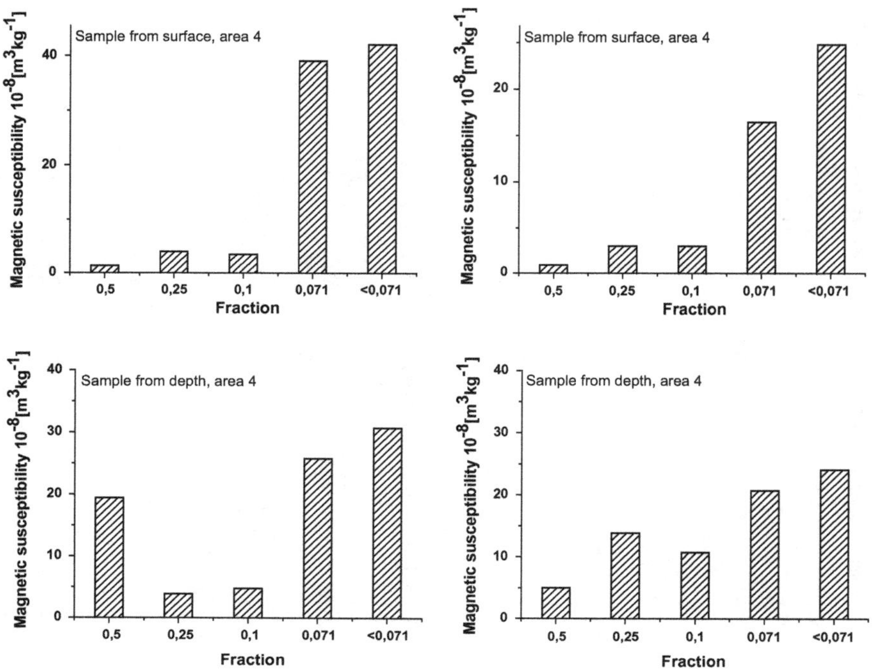

Fig. 3 Distribution of mass magnetic susceptibility (χ) for individual granulometric fractions of surface and depth sediments collected in area 4

Since the fine fractions, with grain size "0.071" and "<0.071", best reflect all the sediment features, detailed analysis is performed for these fractions. Figure 4b presented the distribution of the average values of χ for fine factions of the surface sediments collected from the six areas. In all the areas, the fraction of less than 0.071 mm has higher values of χ_{av} than the fraction of "0.071" indicating more magnetic particles in the finest size of sediments. Distribution of average values of χ together with their standard errors values for fine and finest fractions are presented in Fig. 4c or d. The samples from areas 2 and 3 indicate significant heterogeneity as the deviation of the average values is relatively high. The result may be interpreted by the heterogeneity in the distribution of magnetic particles coming mainly from anthropogenic sources.

The result indicates that most of magnetic particles are present in sediments collected in the city center area. Assuming that the contribution to the magnetic fraction is derived primarily from anthropogenic particles, the high values of magnetic susceptibility indicate higher levels of pollution in these areas.

The χ_{fd} values for fractions "0.071" and "<0.071" are in the range between 0.5 and 3.8%, indicating an insignificant amount of SP particles to magnetic susceptibility (Dearing et al. 1996).

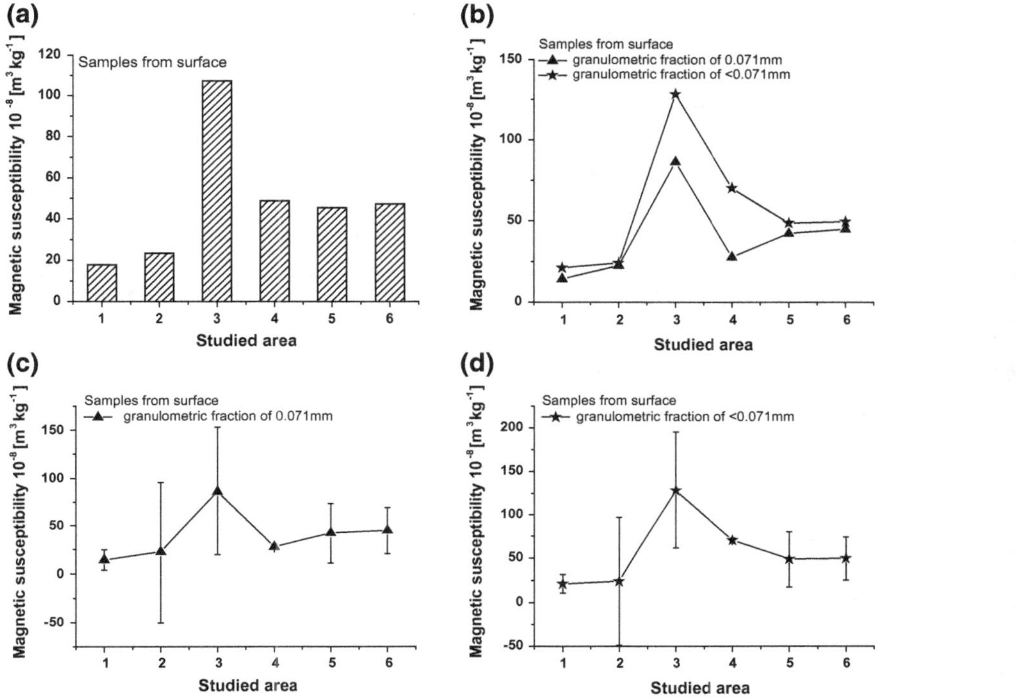

Fig. 4 Distribution of average values of mass magnetic susceptibility (χ) of surface sediments in studied areas (**a**). The average values of χ in particular areas for fine and finest fractions together (**b**) and separately for "0.071" fraction (**c**) and for "<0.071" fraction (**d**). In **c** and **d** graphs, the standard errors of average values of χ are shown

3.3 Mineralogy of Magnetic Fraction

The magnetic mineralogy was determined by estimation of the Curie point on the curve of magnetic susceptibility $\kappa(T)$ versus temperature. Figure 5 shows the representative curves of $\kappa(T)$ for the fine "0.071" and the finest "<0.071" fractions of sediments collected in areas 4 and 5.

During heating, the curves show a very slow increase of κ until about 450 °C and the peak between 400 and 520 °C, which can be due to the Hopkinson peak or formation of new magnetite. The sharp decrease of κ above ~550 °C to near zero is consistent with the Curie temperature of magnetite ~580 °C (Dunlop and Özdemir 1997). During cooling, the curves of $\kappa(T)$ are above the heating ones. The high magnetic enhancement upon the heating is an attribute of formation of a large amount of magnetite. Desenfant et al. (2004) suggests that the significant amount of heating-induced secondary magnetite can be explained by the reduction of organic material.

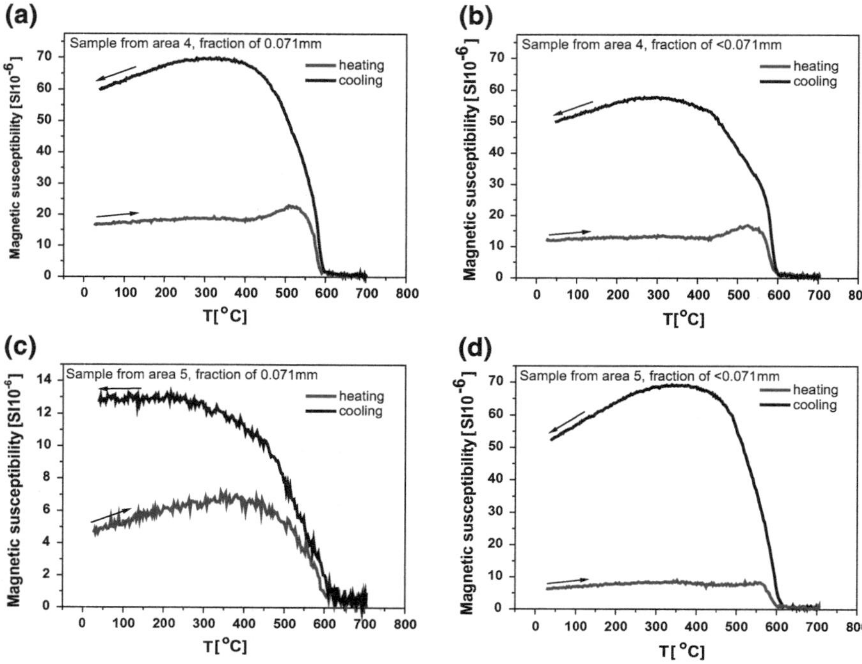

Fig. 5 Curves of magnetic susceptibility versus temperature κ(T) for the fine and finest granulometric fractions of sediments collected from area 4 (**a–b**) and area 5 (**c–d**)

The shape and parameters of hysteresis loops confirm the presence of magnetite as the major magnetic phase in the sediments. Figure 6 shows two hysteresis loops for the sediments collected from areas 3 and 6. The hysteresis loops are relatively narrow and saturated at ~300 mT. The values of hysteresis parameters: H_c from 3 to 8 mT, H_{cr} from 9 to 40 mT, M_s from about 20×10^{-3} A m^2 kg^{-1} to 200×10^{-3} A m^2 kg^{-1}, M_{rs} from about 3×10^{-3} A m^2 kg^{-1} to about 7×10^{-3} A m^2 kg^{-1} indicate the presence of a relatively soft magnetic mineral like multidomain magnetite.

3.4 *The Origin of the Magnetic Particles*

The identification of magnetic particles was provided on magnetic extract of the fine and finest fractions of sediments using an optical microscope. Observations revealed the presence of typical spherules (about 80–90% of the extract) in all the sediments from the six areas (Fig. 7). In particular areas, the variations in the amounts of spherical particles and the differences in their diameter, color and surface morphology were observed. The spherical particles, very often rich in

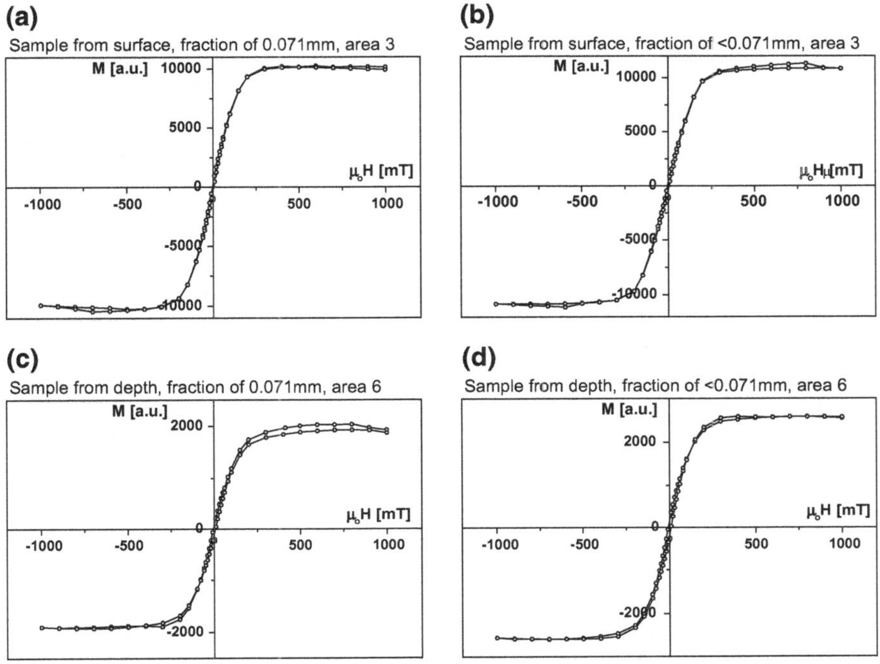

Fig. 6 Hysteresis loops for fine and finest granulometric fractions of sediments collected from area 3 (**a**–**b**) and area 6 (**c**–**d**)

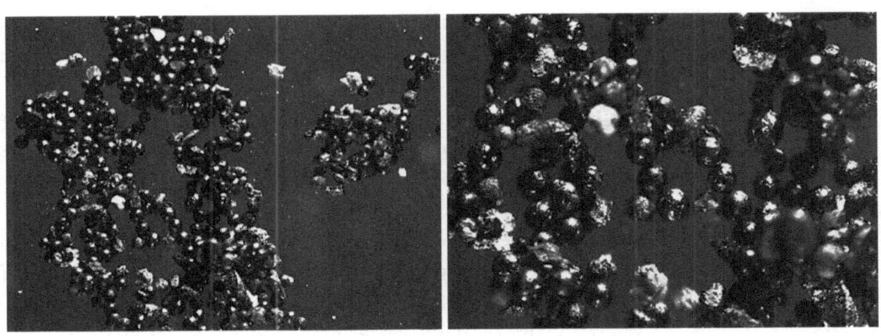

Fig. 7 Optical microscopic images of magnetic particles for the fine granulometric fraction of Vistula river sediments

iron-oxides are a typical product of combustion processes. Since the high concentrations of spherical-shaped particles are observed in the fine and finest fractions, we can infer that the high magnetic susceptibility of these fractions is an evidence of the contribution of the anthropogenic magnetic particles.

The presence of spherical magnetic particles is consistent with the study of polluted river sediments from industrial and urban environments (Jordanova et al. 2004; Li et al. 2011; Zhang et al. 2011; Desenfant et al. 2004). Jordanova et al. (2004) found the significant amount of magnetic spherical particles in the sediments of Danube river close to the big urban area of Bulgaria. The optical images presented in the study revealed the spherules with different surface morphology, structure and chemistry. Most of the particles had complicated dendritic exsolutions within the grain interior. The chemistry of the surface layer showed the presence of Fe-oxides with sulfur and chloride which were found in amounts of one magnitude less for interior grains. Other anthropogenic particles had an orange-peel or dendritic structure of surface or small strongly magnetic spherules adhered to the surface.

Desenfant et al. (2004) reported the anomalies in magnetic susceptibility of sediments collected from Arc river in terms of anthropogenic input. The authors identified the iron-rich spherules derived from the anthropogenic sources such as power plant, city area, and highway bridge.

4 Conclusion

The distribution of magnetic susceptibility of sediments collected from the six areas in the Warsaw part of Vistula river reflects the anthropogenic impact. This allows to estimate the relative level of pollution by the value of magnetic susceptibility of sediments. The results indicate that most of the magnetic particles are present in the sediments collected from the city center area. In the area with relatively low impact of anthropogenic factor, lower values of magnetic susceptibility were recognized.

The study of the granulometric fractions of sediments reveals that the high concentrations of magnetic particles contribute mainly to fine ("0.071") and finest grain fractions ("<0.071").

The thermomagnetic analysis of $\kappa(T)$ curves shows multidomain magnetite as the main magnetic phase in the sediments from the six studied areas. This was deduced from the Curie temperature $\sim$585 °C estimated from $\kappa(T)$ curves which is characteristic for magnetite and from the values of hysteresis loop parameters.

The optical microscopy observations of magnetic extract of sediments confirm that a significant part of magnetic fraction consists of spherical shaped particles of anthropogenic origin.

The application of magnetic method which is very sensitive, inexpensive and fast, allows to trace the temporal and spatial variations of anthropogenic input in the river sediments.

Acknowledgements This study was supported within the project number 3e/IGF PAN/2015 MŁ.

References

Bućko MS, Magiera T, Pesonen LJ, Janus B (2010) Magnetic, geochemical, and microstructural characteristics of road dust on roadsides with different traffic volumes—Case study from Finland. Water Air Soil Pollut. 209(1–4):295–306

Castaneda-Miranda AG, Böhnel HN, Molina-Garza RS, Chaparro MAE (2014) Magnetic evaluation of TSP-filters for air quality monitoring. Atm. Environ. 96:163–174

Chaparro MAE, Bidegain JC, Sinito AM, Gogorza CS, Jurado S (2003) Preliminary results of magnetic measurements on stream sediments from Buenos Aires Province. Argentina. Stud. Geophys. Geod. 47:53–77

Chaparro MAE, Bidegain JC, Sinito AM, Gogorza CS, Jurado S (2004) Magnetic studies applied to different environments (soils and stream sediments) from a relatively polluted area in Buenos Aires Province. Argentina. Environ. Geol. 45:654–664

Dearing JA, Dann RJL, Hay K, Lees JA, Loveland PJ, Maher BA, O'Grady K (1996) Frequency-dependent susceptibility measurements of environmental materials. Geophys. J. Int. 124:228–240

Desenfant F, Petrovský E, Rochette P (2004) Magnetic signature of industrial pollution of stream sediments and correlation with heavy metals: case study from South France. Water Air Soil Pollut. 152:297–312

Dunlop DJ, Özdemir Ö (1997) Rock magnetism: fundamentals and frontiers. Cambridge University Press, Cambridge, p 573

Górka-Kostrubiec B, Jeleńska M, Król E (2014) Magnetic signature of indoor air pollution: household dust study. Acta Geophys. 62:1478–1503

Górka-Kostrbiec B (2015) The Magnetic properties of indoor dust fractions as markers of air pollution inside buildings. Build Environ. 90:186–195

Jeleńska M, Hasso-Agopsowicz A, Kadzialko-Hofmokl M, Kopcewicz B, Sukhorada A, Bondar K, Matviishina Zh (2008) Magnetic structure of the polluted soil profiles from eastern Ukraine. Acta Geophys. 56:1043–1064

Jordanova D, Hoffmann V, Fehr KT (2004) Mineral magnetic characterization of anthropogenic magnetic phases in the Danube river sediments (Bulgarian part). Earth Planetary Sci. Lett. 221:71–89

Knab M, Hoffmann V, Petrovsky E, Kapicka A, Jordanova N, Appel E (2006) Surveying the anthropogenic impact of the Moldau river sediments and nearby soils using magnetic susceptibility. Environ. Geol. 49:527–535

Li F, Li G, Ji J (2011) Increasing magnetic susceptibility of the suspended particles in Yangtze River and possible contribution of fly ash. CATENA 87:141–146

Petrovský E, Kapicka A, Jordanova N, Knab M, Hoffmann V (2000) Low-field magnetic susceptibility: a proxy method of estimating increased pollution of different environmental systems. Environ. Geol. 39:312–318

Sagnotti L, Macri P, Egli R, Mondino M (2006) Magnetic properties of atmospheric particulate matter from automatic air sampler stations in Latinum (Italy): toward a definition of magnetic fingerprints for natural and anthropogenic PM10 sources. J. Geophys. Res. 111:B12S22

Szczepaniak-Wnuk I, Górka-Kostrubiec B (2015) Magnetic particles in indoor dust as marker of pollution emitted by different outside sources. Stud. Geophy. Geod. 60:297–315

Thompson R, Oldfield F (1986) Environmental magnetism. Allen & Unwin, London

Zhang W, Yu L, Lu M, Hutchinson SM, Feng H (2007) Magnetic approach to normalizing heavy metal concentrations for particle size effects in intertidal sediments in the Yangtze Etuary. China. Environ. Pollut. 147:238–244

Zhang Ch, Quiao Q, Piper JDA., Huang B (2011) Assessment of heavy metal pollution from a Fe-smelting plant in urban river sediments using environmental magnetic and geochemical methods. Environ. Pollut. 159:3057–3070

Surface Sediments Pollution Around Small Shipwrecks (Munin and Abille) in the Gulf of Gdańsk: Magnetic and Heavy Metals Study

M. Gwizdała, M. Jeleńska and L. Łęczyński

Abstract Heavy metal contents in sediments from the vicinity of small shipwrecks (Munin and Abille) in the Gulf of Gdańsk and their relationship with mass magnetic susceptibility (χ) were investigated. The values of χ are not very high, however they reveal significant differences between investigated sites. Magnetic susceptibility values vary from 5.33 to 8.02×10^{-8} m^3/kg for Abille and from 3.07×10^{-8} m^3/kg to 12.92×10^{-8} m^3/kg for Munin. It is mainly carried by fraction <0.1 mm in the case of Abille, whereas for Munin χ for <0.1 mm shows different distribution of values. Organic matter content is in the range of 1.60–18.86%, which is characteristic for this area. The concentration of heavy metals is relatively small, as in the case of χ. The relationship amongst analysed heavy metals is strong, but between PLI, LOI and χ is weak. Generally sediments around Munin wreck demonstrate the strong correlation of mass magnetic susceptibility with toxic elements. This is the result of the shipwrecks location and hydrodynamic conditions of this area. The low heavy metal concentration and χ values indicate low level of contamination around both wrecks.

1 Introduction

Sediments are the main store of heavy metals in aquatic environments (Daskalakis and O'Connor 1995). Anthropogenic activities are primary responsible for the presence of toxic substances, such as heavy metals, in the ecosystem (Forstner and Wittman 1979). The occurrence of shipwreck on the seabed is the next source of contamination in the marine environment. The salty water has a negative impact on the steel bode of wreck causing the increase of corrosion rate (Rogowska et al. 2015).

M. Gwizdała (✉) · M. Jeleńska
Institute of Geophysics, Polish Academy of Sciences, Warsaw, Poland
e-mail: mgwizdala@igf.edu.pl

L. Łęczyński
Institute of Oceanography, University of Gdańsk, Gdynia, Poland

M. Jeleńska et al. (eds.), *Magnetometry in Environmental Sciences*,
GeoPlanet: Earth and Planetary Sciences, DOI 10.1007/978-3-319-60213-4_3

Magnetometry is useful tool to study and understand processes occurring in various environments. Thompson and Oldfield (1986), Maher and Thompson (1999), Evans and Heller (2003) discussed and presented reviews of the most important investigations in the field of Environmental Magnetism.

Concentration of heavy metals is connected with the magnetic properties of iron oxides and sulphides present in the pollutions (Jordanova et al. 2012). This relationship is a basis of the use of magnetic techniques in pollution studies. A lot of pollution investigations and magnetic proxies for contamination have been conducted since the 1980s (e.g. Hunt et al. 1984; Beckwith et al. 1986; Ďurža et al. 1993; Strzyszcz and Magiera 1998; Jordanova et al. 2003; Chaparro et al. 2004; Schmidt et al. 2005).

The Gulf of Gdańsk is a shallow water basin with a sandy bottom. It has limited the exchange of water, because of the occurrence of the Hel Peninsula, which separated the Gulf of Gdańsk from Baltic Proper. Within the last few years, in the Gulf of Gdańsk registered the increase of heavy metals concentration in the sediments (Kot-Wasik et al. 2004). This region is one of 132 contamination "hot spots" in the Baltic Sea (Szefer 2002).

In our previous paper (Gwizdała et al. 2016) we investigated the impact of small shipwrecks on marine pollution by use of magnetic methods. The objective of this work was to continue the estimation of environment state in the vicinity of *Abille* and *Munin* shipwrecks, but based on the concentration of toxic elements and mass magnetic susceptibility, and determining the relationship between them.

2 Materials and Methods

2.1 *Study Sites and Sample Handling*

Surface sediment samples were collected from 16 sites in outer Puck Bay, around *Munin* and *Abille* wrecks at the distance of 100 and 200 m (Fig. 1). Sediment collection was performed with the Van Veen grab holder, transported to the laboratory and stored at −20 °C. Before laboratory analyses, samples were dried to constant weight.

2.2 *Methods*

Measurements of mass magnetic susceptibility (χ) have been performed according to the procedure described elsewhere (Gwizdała et al. 2016).

The method of loss on ignition (LOI) is based on measuring loss of sample mass after ignition of organic matter. The sediment samples of known mass were heated at 550 °C for 1 h, then cooled and re-weighed. The content of organic matter is

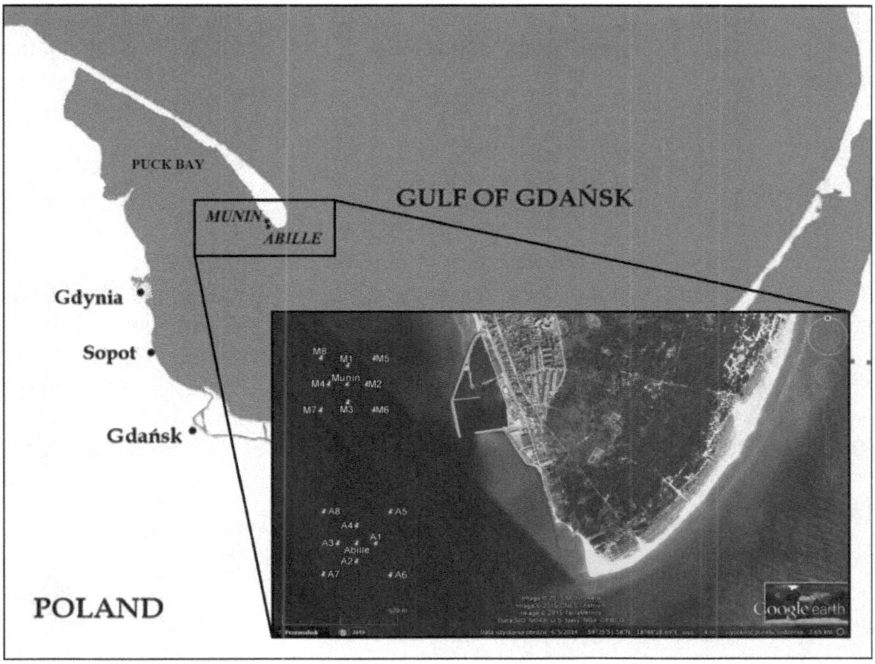

Fig. 1 Localisation of *Munin* and *Abille* wrecks and position of sediment sampling in their vicinity

expressed as a percentage difference between the initial and final sample mass divided by the initial sample mass.

The contents of 20 chemical elements (Al, As, Ba, Be, Cd, Co, Cr, Mn, Mg, Mo, Ni, P, Pb, Sr, Ti, Tl, U, Zn, Cu, Fe) in the sediment samples were measured using the Inductively Coupled Plasma Mass Spectrometry (ICP-MS). Before measurement, the sediments of silty-clay fraction (<0.1 mm) were mineralized using the concentrated nitric acid (in a proportion of 0.1 g of sediment in 5 ml of acid). The total concentrations were determined in ppm of dried mass (mg/kg).

2.3 *PLI*

The correlation between magnetic properties and heavy metal concentration was determined by the Tomlinson Pollution Load Index (PLI) (Tomlinson et al. 1980; Angulo 1996). The definition of PLI given by Tomlinson et al. (1980) is:

$$\sqrt[n]{CF_1 * CF_2 * \ldots * CF_n}$$

where CF is the quotient the concentration of each metal and the value of background adopting the minimal concentration value for this metal, and n is the number of these metals.

3 Results

Preliminary magnetic results of investigated samples were presented in our previous paper (Gwizdała et al. 2016). These results revealed a relatively small values of the magnetic susceptibility, however, they showed significant differences between studied sites. The *Abille* set of samples reached values of χ between 5 and 8 $\times$ 10^{-8} m^3/kg. In the vicinity of *Munin* wreck results of χ were more varied ($3 < \chi < 13 \times 10^{-8}$ m^3/kg). The values of χ increased towards SWW around *Munin* wreck and towards SW close to *Abille* wreck (Fig. 2) (Gwizdała et al. 2016).

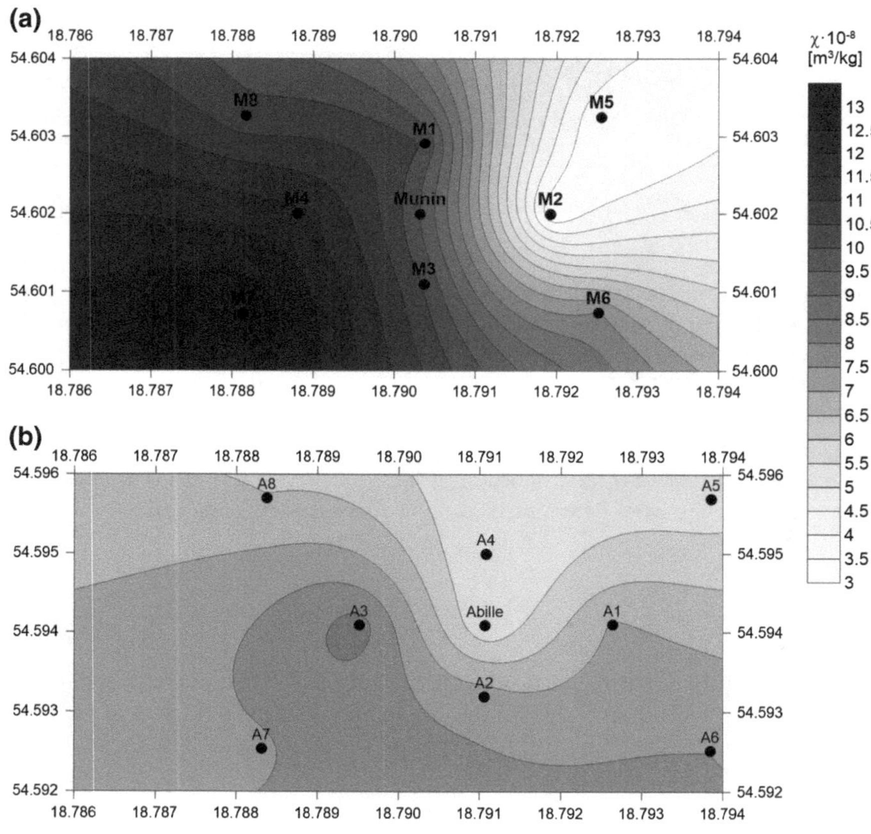

Fig. 2 Spatial distribution of magnetic susceptibility (χ) around **a** *Munin* and **b** *Abille* wrecks (after Gwizdała et al. 2016)

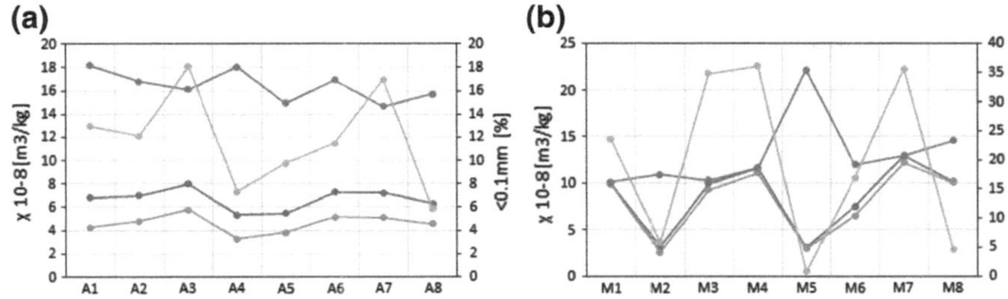

Fig. 3 On the main axis is the mass magnetic susceptibility for whole sample (*blue line*), coarse grained fraction (*grey line*) and fine grained fraction (<0.1 mm) (*orange line*). On the secondary axis is the percentage of fraction <0.1 mm in whole sample (*yellow line*). Results are presented for *Abille* (**a**) and *Munin* (**b**) sets

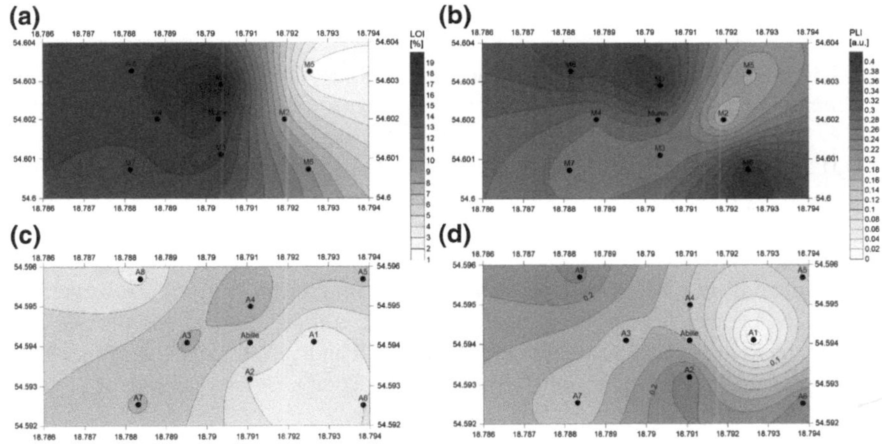

Fig. 4 Spatial distribution of LOI (**a**, **c**) and PLI (**b**, **d**) around shipwrecks

Besides the measurements of χ for whole sample, we have performed the measurement of silty-clay fraction (<0.1 mm). The values of magnetic susceptibility for fraction <0.1 mm were higher than those for the whole sample (Fig. 3). In point where the differences were significant, the percentage of the finest fraction in whole specimen is low (<7%). This analysis showed that the main carrier of χ is the fine grained fraction in the case of *Abille* (Fig. 3a), whereas for *Munin* magnetic susceptibility for <0.1 mm shows different distribution of values (Fig. 3b).

The lost on ignition (LOI) values are in the range of 1.60–18.86% for *Munin* set. The LOI in these samples is higher and more diversified. Whereas specimens around *Abille* wreck are clustered between 3 and 8%. The highest percent of LOI extends through the middle of the *Abille* area towards SW (Fig. 4c). Whereas, the highest LOI values (>14%) for *Munin* set occurs in the middle and west part of investigated area (Fig. 4a).

Table 1 shows concentration of 20 toxic elements in investigated samples. The content of heavy metals is relatively small. Higher values are characteristic for *Munin* group of samples than for *Abille* one. The PLI coefficient shows, that the lowest values extend from NE to SW in the vicinity of both wrecks (Fig. 4b and d). Table 2 presents the correlation coefficients (r) between investigated elements for all samples. Correlation coefficient amongst analysed heavy metals is very high, only relation between Be and Mo with other elements is relatively small ($r_{average} = 0.65$ and $r_{average} = 0.66$, respectively). The lowest r is for the pair Be/Mo (0.35), Be/U (0.46) and Mo/P (0.4), and the highest between Cd, Cu and Zn ($r \approx 0.99$). Samples from *Munin* wreck have a characteristic linear distribution of correlation between heavy metals (Fig. 5). While the *Abille* set of samples, can be divided into four types of association. In the most cases, this set is strict clustered at low values (Fig. 5a). In the second type it is mixed with specimens from *Munin* wreck, as a linear function (Fig. 5b). In the next type, the *Abille* set is separated as independent linear correlation (Fig. 5c) but in the last one, lack of linear relationship is caused by constant values of one element (Fig. 5d).

The association between magnetic susceptibility and chemical elements in sediments around *Munin* wreck is not so strong as in other cases, but it occurs (Table 3). Correlation coefficient reaches the highest values for Fe and Tl ($r = 0.693$). Also the correlation between χ and As, Cd, Co, Mn, Ni, Sr, Zn, Cu is relatively high ($0.54 < r < 0.62$). The values below 0.16 occur for Mo, Pb and Ti. In the case of *Abille* set, this relationship is very weak. Only Cd and Zn indicate the association with χ ($r > 0.5$).

The dependence between PLI, LOI and χ is very weak (Fig. 6). However, the difference between investigated sites occurs. The *Abille* group of samples is clustered in the lower part of studied range of PLI, LOI and χ. Whereas, the Munin set is shifted towards higher values of PLI, χ and LOI, except for two samples (M2, M5).

4 Discussion

The high correlation coefficient between analysed trace elements ($r > 0.5$, except 3 cases) indicates the same source. It means that the main source of heavy metals in this area is the presence of shipwrecks.

The relationship of magnetic susceptibility with chemical elements, mainly occurs in deposits collected around *Munin* wreck. Additionally, the concentration of heavy metals and value of mass magnetic susceptibility are also higher in *Munin* group of samples. It could be the result of the localisation of shipwrecks and hydrodynamic conditions of this area. Wreck of *Munin* is located inside the bay in contrast to *Abille* shipwreck situated close to the headland of Hel (Fig. 1), and exposed on currents flow from the open sea. In this position, the sediments and contamination included therein washout and redeposited into the Baltic Sea resulting in low heavy metal contents and χ values.

Table 1 Concentration of chemical elements in sediments around *Munin* (M1-M8) and *Abille* (A1-A8) wrecks

Sample	Al	As	Ba	Be	Cd	Co	Cr	Cu	Fe	Mn	Mg	Mo	Ni	P	Pb	Sr	Ti	Tl	U	Zn
M1	3700.9	9.8	35.2	0.4	1.8	4.5	28.1	31.1	16,210.7	167.1	6887.7	1.4	23.6	664.8	56.1	45.7	84.6	0.3	1.5	116.5
M2	1215.2	5.1	12.5	0.2	0.6	1.8	6.8	12.3	5775.4	62.2	2424.9	0.9	10.8	338.2	23.6	21.8	24.1	0.1	0.9	58.3
M3	2056.8	7.2	18.4	0.4	1.1	2.9	13.8	20.8	10,961.0	118.1	4384.0	0.7	18.2	514.5	26.2	31.0	36.1	0.2	0.9	84.0
M4	2205.1	7.2	19.8	0.4	1.3	3.0	15.6	23.1	11,225.9	115.1	4348.4	1.0	17.4	503.9	29.9	28.5	38.6	0.2	1.1	90.5
M5	1490.6	5.7	19.7	0.2	0.6	1.7	10.1	15.2	4880.9	49.4	2834.3	0.9	10.0	524.1	32.1	22.8	47.6	0.1	0.9	50.9
M6	4532.0	8.4	31.4	1.0	1.8	4.0	29.8	31.8	13,956.4	167.4	8233.7	1.1	25.0	845.1	51.6	40.6	102.1	0.3	1.2	121.1
M7	2170.9	7.1	19.8	0.3	1.0	3.0	14.1	22.6	10,833.8	106.7	4513.7	0.8	17.7	568.3	26.5	33.4	41.2	0.2	1.1	83.7
M8	3207.2	8.9	33.8	0.4	1.6	4.2	24.9	28.3	14,199.2	140.7	6247.9	1.3	22.2	584.5	52.7	43.5	82.1	0.3	1.3	102.7
Average	2572.4	7.4	23.8	0.4	1.2	3.1	17.9	23.1	11,005.4	115.8	4984.3	1.0	18.1	567.9	37.3	33.4	57.0	0.2	1.1	88.5
Sample	Al	As	Ba	Be	Cd	Co	Cr	Cu	Fe	Mn	Mg	Mo	Ni	P	Pb	Sr	Ti	Tl	U	Zn
A1	2120.9	6.3	23.3	0.0	0.5	2.5	10.6	11.3	7500.4	106.5	2833.2	0.6	12.6	554.9	38.2	28.3	54.4	0.2	0.9	46.6
A2	2448.5	5.8	26.4	0.4	0.5	2.9	12.3	13.5	8956.5	124.7	5739.9	1.1	16.6	619.1	38.8	26.9	63.3	0.1	0.9	44.9
A3	1274.1	4.3	15.8	0.2	0.5	1.9	6.4	10.8	7095.6	88.0	3510.3	0.5	10.8	380.3	13.4	18.9	27.2	0.1	0.6	44.5
A4	1242.5	4.4	15.0	0.2	0.3	1.8	5.9	9.7	5859.2	89.4	3590.4	0.5	10.3	531.0	13.1	21.2	25.6	0.1	0.7	39.3
A5	1137.7	4.2	12.5	0.3	0.4	1.7	5.4	9.9	5486.8	72.1	3096.2	0.6	11.0	428.2	13.7	21.5	23.4	0.1	0.7	39.3
A6	2256.8	5.6	24.3	0.6	0.5	2.8	11.4	13.4	9947.4	154.9	5393.7	0.7	15.8	609.2	34.5	29.8	68.9	0.1	0.9	50.4
A7	1185.6	4.2	13.4	0.2	0.4	1.9	6.1	10.7	6374.7	81.6	3234.5	0.7	10.4	355.6	14.0	17.7	23.8	0.1	0.7	44.0
A8	2363.1	5.9	24.5	0.2	0.5	3.1	10.2	13.5	9276.2	113.6	5783.3	1.5	15.0	469.8	25.5	26.6	63.4	0.1	1.1	49.5
Average	1753.7	5.1	19.4	0.3	0.5	2.3	8.5	11.6	7562.1	103.8	4147.7	0.8	12.8	493.5	23.9	23.9	43.7	0.1	0.8	44.8

Table 2 Correlation coefficients (r) between chemical elements in sediments around Munin and Abille wrecks

	Al	As	Ba	Be	Cd	Co	Cr	Mn	Mg	Mo	Ni	P	Pb	Sr	Ti	Tl	U	Zn	Cu	Fe
Al	1.00	0.87	0.92	0.75	0.85	0.93	0.96	0.88	0.93	0.69	0.94	0.88	0.89	0.91	0.94	0.89	0.84	0.83	0.86	0.89
As		1.00	0.85	0.51	0.94	0.93	0.95	0.74	0.73	0.69	0.92	0.70	0.89	0.96	0.77	0.98	0.93	0.93	0.95	0.93
Ba			1.00	0.53	0.74	0.92	0.88	0.84	0.86	0.73	0.85	0.80	0.94	0.89	0.96	0.82	0.82	0.68	0.75	0.85
Be				1.00	0.64	0.62	0.69	0.75	0.81	0.35	0.75	0.75	0.53	0.59	0.67	0.57	0.46	0.65	0.66	0.66
Cd					1.00	0.88	0.95	0.70	0.74	0.63	0.92	0.65	0.78	0.91	0.68	0.95	0.87	**0.99**	**0.99**	0.92
Co						1.00	0.93	0.90	0.89	0.75	0.97	0.73	0.86	0.96	0.86	0.92	0.89	0.85	0.88	0.98
Cr							1.00	0.81	0.85	0.67	0.95	0.81	0.90	0.95	0.86	0.95	0.88	0.93	0.96	0.93
Mn								1.00	0.90	0.51	0.89	0.79	0.75	0.83	0.83	0.75	0.67	0.68	0.70	0.90
Mg									1.00	0.70	0.90	0.81	0.74	0.80	0.89	0.75	0.72	0.71	0.76	0.86
Mo										1.00	0.66	0.40	0.68	0.64	0.70	0.66	0.85	0.58	0.63	0.66
Ni											1.00	0.79	0.83	0.95	0.82	0.93	0.85	0.91	0.93	0.97
P												1.00	0.79	0.77	0.86	0.72	0.61	0.64	0.70	0.70
Pb													1.00	0.90	0.92	0.84	0.84	0.74	0.79	0.80
Sr														1.00	0.83	0.94	0.89	0.89	0.91	0.94
Ti															1.00	0.75	0.77	0.65	0.71	0.77
Tl																1.00	0.92	0.94	0.96	0.93
U																	1.00	0.84	0.87	0.86
Zn																		1.00	**0.99**	0.90
Cu																			1.00	0.91
Fe																				1.00
$r_{average}$	0.88	0.86	0.83	0.65	0.84	0.88	**0.89**	0.79	0.82	0.66	0.89	0.74	0.82	0.87	0.81	0.86	0.82	0.82	0.85	0.87

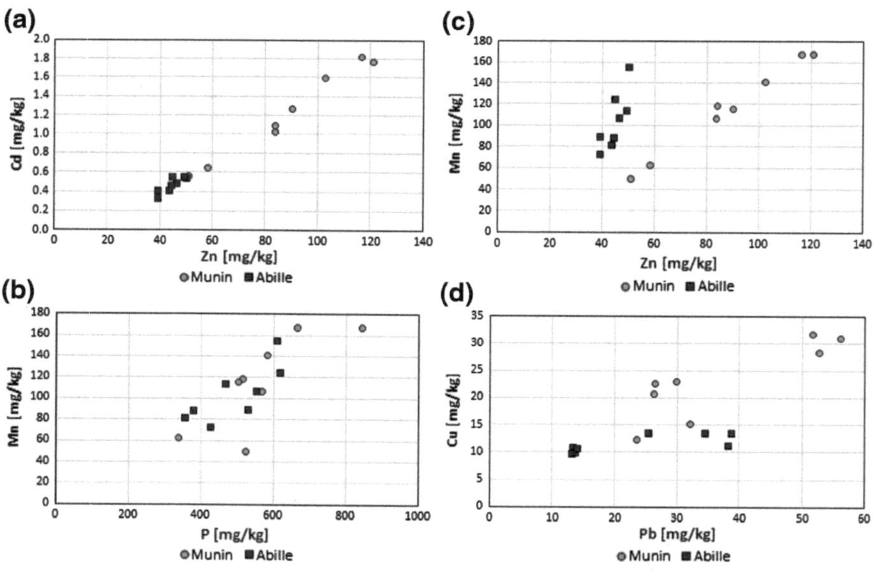

Fig. 5 Comparison of selected chemical elements in *Munin* and *Abille* set of samples

Table 3 Correlation coefficients (r) between mass magnetic susceptibility and chemical elements in sediments around *Munin* and *Abille* wrecks

	Al	As	Ba	Be	Cd	Co	Cr	Mn	Mg	Mo	Ni	P	Pb	Sr	Ti	Tl	U	Zn	Cu	Fe
Munin set																				
χ	0.351	**0.595**	0.328	0.261	**0.546**	**0.613**	0.392	**0.601**	0.432	0.152	**0.613**	0.294	0.156	**0.548**	0.15	**0.693**	0.475	**0.583**	**0.587**	**0.693**
Abille set																				
χ	0.223	0.175	0.277	0.159	**0.501**	0.265	0.319	0.337	0.14	0.026	0.23	−0.11	0.269	0.025	0.244	0.162	-0.07	**0.573**	0.39	0.445

Fig. 6 Dependence between PLI, LOI and χ in *Munin* and *Abille* set of samples

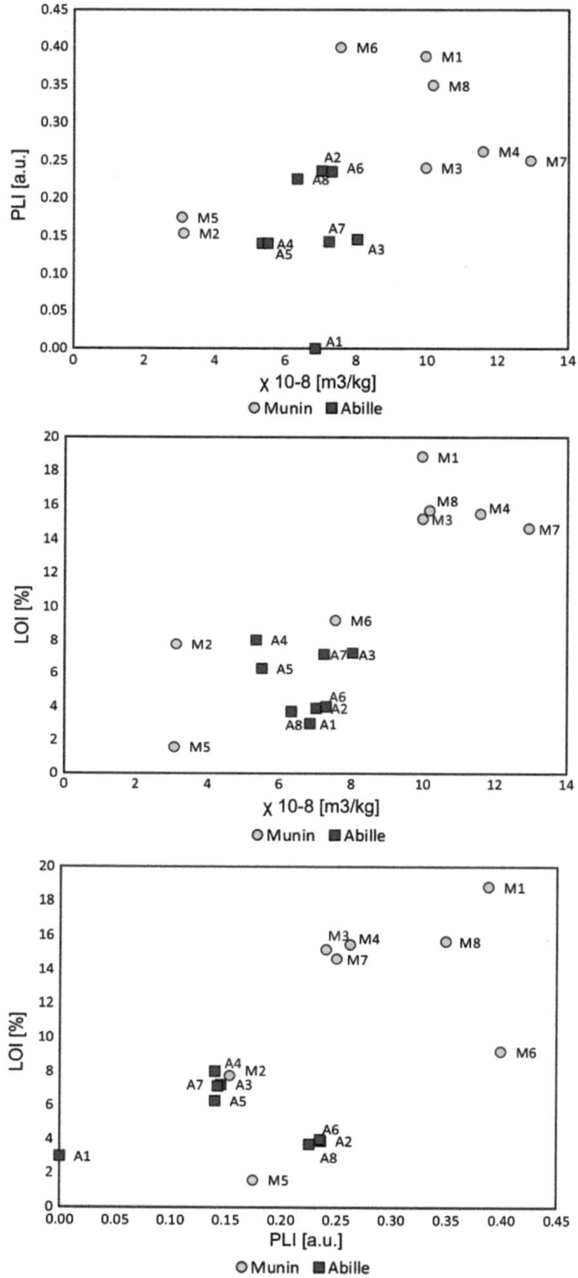

According to Gwizdała et al. (2016) the main magnetic carrier in investigated sediments is magnetite with small amount of maghemite and hematite in *Abille* set and one mineral—magnetite or maghemite in *Munin* set. The presence of these iron oxides increases the correlation coefficient between Fe and χ.

Poor correlation of magnetic susceptibility with the Tomlinson Pollution Load Index could be a result of low χ values. According to Schmidt et al. (2005), strong association only exists for specimens of enhanced magnetic susceptibility. In their study strong correlation between χ and Fe occurs for soils with values exceeding 176×10^{-8} m^3/kg. Similar remarks can be applied to the study of Ďurža et al. (1993) who detected only poor correlation between χ and heavy metal contents, because their enhancement of magnetic susceptibility across the investigated site was weak. Also Beckwith et al. (1986) found that the association of magnetic susceptibility with heavy metals is lost when specimens with low values of χ were added to the analysis. Whereas Chaparro et al. (2004) explained non-significant correlation between heavy metals and magnetic susceptibility taking into account the discrimination of the two distinguishing groups of sediments.

The linkage between magnetic parameters and pollutant contents have been provided in many papers, for example Hunt et al. (1984), Beckwith et al. (1986), Jordanova et al. (2003), Chaparro et al. (2004). This relationship depends on the origin of the study material. Strzyszcz (1993) found a significant correlation between magnetic susceptibility and concentration of heavy metals in soils influenced by industrial emissions. Beckwith et al. (1986) observed this strong relationship for urban source sediments. High correlation coefficient usually occurs when investigated material is anthropogenic origin. The industrial and urban process composed of the additional influx of magnetic particles, including heavy metals (Chaparro et al. 2006). The low values of χ, toxic element contents and weak correlation coefficient PLI/χ can be a result of low level of anthropogenic pollution. *Munin* and *Abille* wrecks sunk in 1945, have probably too small corrosion rate and theirs influence on the seabed is not very high.

The trace element concentrations obtained for sediments around *Munin* and *Abille* wrecks in contrast to Szefer et al. (2009) results for Gulf of Gdańsk, are very low. Average values are from 1.7 to even 119.1 times lower for samples collected close to wrecks than for those from Gulf of Gdańsk, except for one element (Mo), which reaches a higher value around wrecks (Table 4). This is the another confirmation of low level of contamination in the vicinity of *Munin* and *Abille* wrecks.

The low LOI values are characteristic for this area. Huzarska (2013) and Brodecka et al. (2013) recorded the similar quantity of organic matter in outer Puck Bay.

Table 4 Comparison of selected element contents between sediments around both wrecks and Gulf of Gdańsk

Localisation	Al	As	Be	Cd	Co	Cr	Cu	Fe	Mn	Mo	Pb	Sr	Ti	Zn
Munin and Abille wrecks	2163	6.3	0.3	0.8	2.7	13.2	17.4	9283.8	109.8	0.9	30.6	28.6	50.4	66.6
Gulf of Gdańsk (Szefer et al. 2009)	50,000	12	1.5	2	9	90	35	37,000	465	0.4	52.0	103.0	6000	132

5 Conclusions

Sixteen surface sediment samples collected in the vicinity of two shipwrecks have been subject to heavy metal contents and magnetic analysis. The results show that the concentration of trace elements and values of mass magnetic susceptibility are relatively small. In most cases, correlation coefficient among analysed chemical elements is high which means, that the source of heavy metals in the study area is one. The strong relationship of mass magnetic susceptibility with toxic elements, mainly occurs in sediments around *Munin* wreck, the strongest is for thallium and iron. Whereas, the association between PLI, LOI and χ is weak. Poor correlation of magnetic susceptibility with PLI could be an effect of weak enhancement of χ across the studied specimens. Despite the fact that obtained results indicate low level of pollution around both wrecks, the difference among investigated sites occurs. The *Munin* group of samples contains higher and more diversified values of results of conducted analysis in comparing to *Abille* one.

Acknowledgements The publication has been partially financed from the funds of the Leading National Research Centre (KNOW) received by the Centre for Polar Studies for the period 2014–2018. The authors would like to thank Dr. P. Szwarczewski for IPC-MS and LOI analyses.

References

Angulo E (1996) The Tomlinson Pollution Load Index applied to heavy metal, ‘Mussel-Watch’ data: a useful index to assess coastal pollution. Sci. Total Environ. 187(1):19–56. doi:10.1016/0048-9697(96)05128-5

Beckwith PR, Ellis JB, Revitt DM, Oldfield F (1986) Heavy metal and magnetic relationships for urban source sediments. Phys. Earth Planet. Inter. 42:67–75

Brodecka A, Majewski P, Bolałek J, Klusek Z (2013) Geochemical and acoustic evidence for the occurrence of methane in sediments of the Polish sector of the southern Baltic Sea. Oceanologia 55(4):951–978. doi:10.5697/oc.55-4.951

Chaparro MAE, Bidegain JC, Sinito AM, Jurado S, Gogorza CS (2004) Relevant magnetic parameters and heavy metals from relatively polluted stream-sediments—spatial distribution along a Cross city Stream in Buenos Aires Province, Argentina. Stud. Geophys. Geod. 48 (3):615–636

Chaparro MAE, Gogorza CSG, Chaparro MAE, Irurzun MA, Sinito AM (2006) Review of magnetism and heavy metal pollution studies of various environments in Argentina. Earth Planets Space 58:1411–1422

Daskalakis KD, O'Connor TP (1995) Distribution of chemical concentrations in US coastal and estuarine sediment. Marine Environ. Res. 40:381–398

Ďurža O, Gregor T, Antalova S (1993) The effect of the heavy metals soil contamination on the magnetic susceptibility. Acta Universitatis Carolinae Geologica 37:135–143

Evans ME, Heller F (2003) Environmental magnetism, Principles and applications of enviromagnetics. Academic Press, San Diego

Forstner U, Wittman GTW (1979) Metal pollution in the aquatic environment. Springer-Verlag, Berlin, Heidelberg, New York, pp 1–486

Gwizdała M., Jeleńska M, Łęczyński L (2016) Magnetometry as a Tool to Estimate the Pollution of Marine Environment Around Small Shipwrecks (Gulf of Gdańsk)–Preliminary Results. Acta Geophysica. 64(5):1691–1702. doi:10.1515/acgeo-2016-0056

Hunt A, Jones J, Oldfield F (1984) Magnetic measurements and heavy metals in atmospheric particulates of anthropogenic origin. Sci. Total Environ. 33:129–139

Huzarska K (2013) Spatial distribution of biological and physical sediment parameters in the western Gulf of Gdańsk. Oceanologia 55(2):453–470. doi:10.5697/oc.55-2.453

Jordanova D, Jordanova N, Lanos P, Petrov P, Tsacheva T (2012) Magnetism of outdoor and indoor settled dust and its utilization as a tool for revealing the effect of elevated particulate air pollution on cardiovascular mortality. Geochem. Geophys. Geosyst. 13(8):Q08Z49. doi:10.1029/2012GC004160

Jordanova NV, Jordanova DV, Veneva L, Andorova K, Petrovsky E (2003) Magnetic response of soils and vegetation to heavy metal pollution—a case of study. Environ. Sci. Tech. 37:4417–4424

Kot-Wasik A, Dębska J, Namieśnik J (2004) Monitoring of organic pollutants in coastal waters of the Gulf of Gdańsk. Southern Baltic, Marine Pollution Bullet. 49(3):264–276

Maher BA, Thompson R (1999) Quaternary climate, environments and magnetism. Cambridge University Press, Cambridge

Rogowska J, Kudłak B, Tsakovski S, Gałuszka A, Bajger-Nowak G, Simeonov V, Konieczka P, Wolska L, Namieśnik J (2015) Surface sediments pollution due to shipwreck s/s "Stuttgart": a multidisciplinary approach. Stoch. Environ. Res. Risk Assess. 29:7. doi:10.1007/s00477-015-1054-0

Schmidt A, Yarnolda R, Hilla M, Ashmorea M (2005) Magnetic susceptibility as proxy for heavy metal pollution: a site study. J. Geochem. Explor. 85:109–117. doi:10.1016/j.gexplo.2004.12.001

Strzyszcz Z (1993) Magnetic susceptibility of soils in the areas influenced by industrial emssions. Soil monitoring. Biddgtuser Verlag Basel, Monte Verita, pp 155–269

Strzyszcz Z, Magiera T (1998) Magnetic susceptibility and heavy metal contamination in soils of southern Poland. Phys. Chem. Earth 23:1127–1131

Szefer P (2002) Metals, metalloids and radionuclides in the Baltic Sea ecosystem, trace metals in the environment. Elsevier, London

Szefer P, Glasby GP, Geldon J, Renner RM, Björn E, Snell J, Frech W, Warzocha J (2009) Heavy-metal pollution of sediments from the Polish exclusive economic zone, southern Baltic Sea. Environ. Geol. 57:847–862. doi:10.1007/s00254-008-1364-3

Thompson R, Oldfield F (1986) Environmental magnetism. Allen & Unwin, London

Tomlinson DC, Wilson JG, Harris CR, Jeffrey DW (1980) Problems in the assessment of heavy metals in estuaries and the formation pollution index. Helgol. Mar. Res. 33:566–575

From Deserts to Glaciers: Magnetometry in Paleoenvironmental Studies in Central Asia

Monika Mętrak, Fabian Welc, Piotr Szwarczewski and Małgorzata Suska-Malawska

Abstract Central Asia is a vast area extending from the Caspian Sea to the western borders of China, with highly differentiated geography (high mountains, excessive deserts, grassy steppes), geology and anthropological history. As such it poses a perfect site for various studies, including paleoenvironmental and paleoclimatic research. Up to now, we performed two distinct studies in Central Asia: (1) a preliminary study on the Holocene loess-soil succession in the Karasu Valley, at the Tien Shan foothills in Uzbekistan; (2) and a study on development of high mountain lakes in the Pamir, based on lake sediments of the Rangkul Lake in the Eastern Pamir, Tajikistan (initial results presented here). In both studies we used magnetic properties of studied sediments and soils as one of parameters in multi-proxy analyses. Paleosols from the Karasu Valley developed into continuous and uninterrupted sedimentary sequences interbedded with loess horizons. These sequences are characterized by diversified magnetic susceptibility values that reflect changes in their formation. Sediments form the Rangkul Lake show 5 significant shifts in deposition processes, caused by changes in water level. These shifts are clearly reflected both in sediment composition and in magnetic susceptibility of deposited material.

Keywords Central Asia · Paleoenvironment · Magnetic susceptibility

M. Mętrak (✉) · M. Suska-Malawska
Faculty of Biology, Biological and Chemical Research Centre, University of Warsaw, Żwirki i Wigury 101, 02-089 Warsaw, Poland
e-mail: mmetrak@biol.uw.edu.pl

F. Welc
Institute of Archaeology, Cardinal Wyszyński University in Warsaw, Wóycickiego 1/3, 01-938 Warsaw, Poland

P. Szwarczewski
Faculty of Geography and Regional Studies, University of Warsaw, Krakowskie Przedmieście 30, 00-927 Warsaw, Poland

M. Jeleńska et al. (eds.), *Magnetometry in Environmental Sciences*,
GeoPlanet: Earth and Planetary Sciences, DOI 10.1007/978-3-319-60213-4_4

1 Introduction

Though there is no universal definition of Central Asia, the existing ones describe this area as a core region of the Eurasian continent, extending from the Caspian Sea to the western borders of China (Fig. 1a). Traditionally, this region consists of five former Soviet republics—Kazakhstan, Uzbekistan, Tajikistan, Kyrgyzstan and Turkmenistan, sometimes also Afghanistan is included (Zhang et al. 2013; Encyclopaedia Britannica 2016). Geography of this vast area is highly differentiated, including high mountains (Tien Shan, Pamir, Altai), excessive desserts, occupying c.a. 60% of the region (Gobi, Karakum, Kyzylkum), and treeless grassy steppes. Under arid and semi-arid continental climate, these areas are mostly unsuitable for farming, except along the margins of the Amu Darya and Syr Darya river systems. Therefore, irrigation agriculture is dominating in Central Asia, leading to serious man-made environmental transformations (e.g. secondary salinization, lowering water level in endorheic lakes, such as the Aral Lake or the Balkhash Lake) (Zhang et al. 2013; Encyclopaedia Britannica 2016).

Interestingly, this extremely dry area is located in the temperate zone where it is normally occupied by the westerly humid climate. However, the Central Asian inland is rain shadowed by young mountainous formations emerging from the collision between the Indian and the Eurasian tectonic plates (i.e. the Himalayas Hindu Kush, Pamir, Tibetan Plateau) (Dambricourt and Gaillard 2011; Zhang et al. 2013). Intense aridification of the Asian inland happened in the Cenozoic and was one of the most dominant climate changes on earth. This transition may be linked to (1) the retreat of the Tethys Sea, (2) the late Miocene-Pliocene rapid uplift of the Tibetan Plateau, the Kunlun Shan and the Pamir, and (3) the long term global cooling due to the Arctic ice sheet expanding since the late Miocene, which caused decrease in moisture of monsoon and westerlies, and thus reduced precipitation (Wang et al. 2014; Zhang et al. 2013).

As the rained-shadowed Central Asian valleys were extremely arid, human settlements spread over the fringes of the Western Himalayas, as early as 2 Ma ago, when both the Himalayas and the Tibet Plateau were 2700 m below the present altitude. Human settlements thrived there over periods of warmer/wetter climate and their inhabitants relayed on high plateaus as their summer hunting grounds. Around 1 Ma ago an important uplift of mountain ranges occurred and a severe global cooling down followed, which disturbed the existing ecosystems. Yet, hunting communities existed at the foothills of the Western Himalayas till around 5000 BP, when a shift to pastoral and agricultural economy took place (Dambricourt and Gaillard 2011).

The oldest human settlements on the Central Asian lowlands were dated at around 130,000 BP (Sel'Ungur cave in Ferghana region, Uzbekistan). Interestingly, there are very few sites dated at the Late Paleolithic, probably due to unfavorable environmental conditions that hampered human settlement (Ranov and Davis 1979; Islamov et al. 1988; Welc et al. 2016). Unfortunately, correlations between climate changes and Paleolithic human cultures in Central Asia has remained so far insufficiently

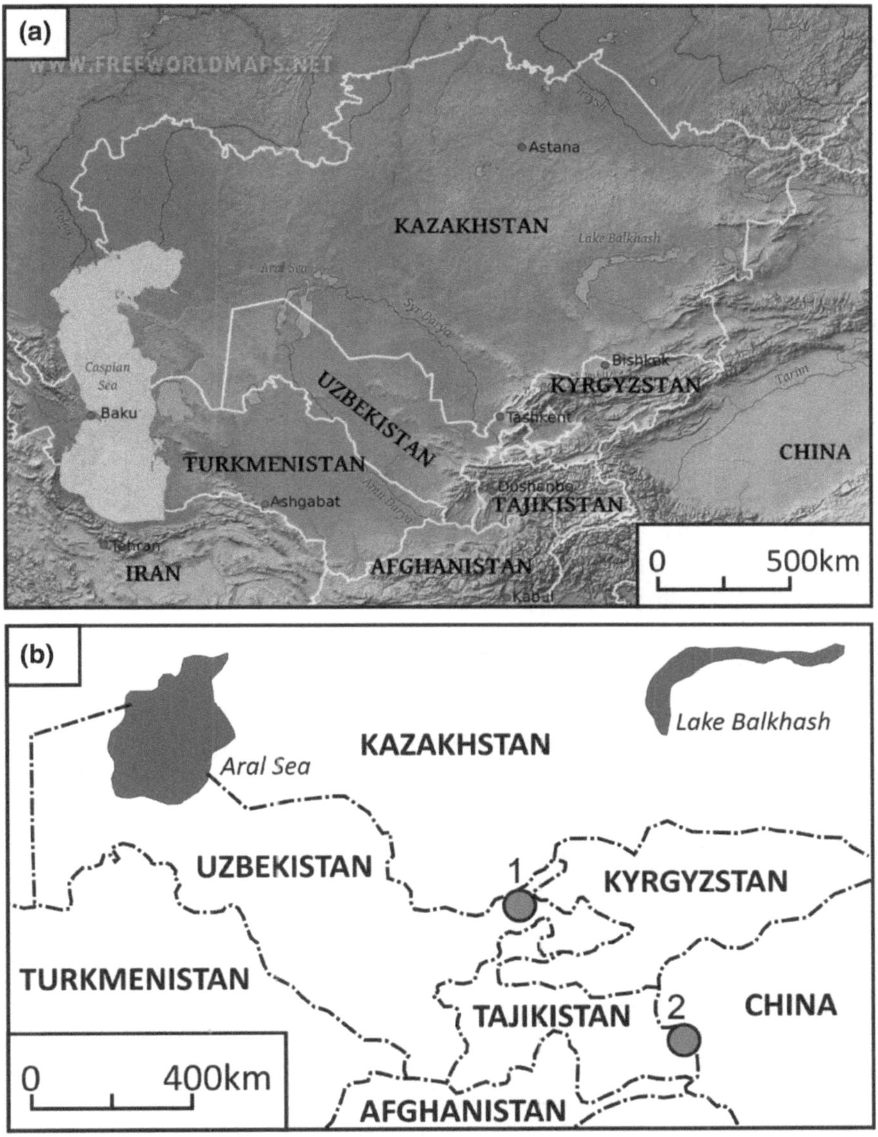

Fig. 1 Map of Central Asia (**a**, freeworldmaps.net, modified) with location of described research sites (**b**). *1* Holocene loess-soil succession, Karasu Valley, Uzbekistan, *2* Rangkul Lake, Pamir, Tajikistan

studied (Glantz et al. 2004; Welc et al. 2016). At the beginning of the Holocene, with the onset of interglacial conditions, new habitats were open for human population. Warm and humid climate in the Early Boreal (BO-1, ca. 9200–8000 BP) coincided with the beginning of the Neolithic in Kazakhstan and Uzbekistan, and was followed by a subsequent cooling in the middle of the Atlantic (AT-2, ca. 7000 BP). This severe shift in climatic conditions caused extinction of large steppe animals. As an adaptation, humans began to domesticate horses and cattle (Baibatsha 2012). The most favorable conditions for human settlement occurred in Uzbekistan and Kazakhstan in the Late Atlantic (AT-3, ca. 6000–5000 BP), when warm and humid climate promoted vegetation, that in turn stimulated cattle breeding (Baibatsha 2012). Around 4200 BP, after one of the most significant Holocene global climate fluctuations (Bond Event No. 3), thriving Central Asian steppes began to rapidly dry up, resulting in decline in cattle population, famine and human migrations (Baibatsha 2012). This anomaly lasted around 300 years and brought about rapid collapse of Bronze Age civilizations worldwide.

With differentiated and intricate history, both in geological and anthropological context, Central Asian countries offer many opportunities of scientific studies concerning paleoclimate and its impact on human development. Up to now, we performed two distinct studies in Central Asia (Fig. 1b), with the use of two main sources of paleoenvironmental data: (1) a preliminary study on the Holocene loess-soil succession at the Tien Shan foothills in Uzbekistan (Fig. 2a); (2) and a study on development of high mountain lakes in the Pamir, based on lake sediments composition (Fig. 2b). This article presents short summary of our recent research in "the heart of Asia".

2 The Deserts: Holocene Loess-Soil Succession in the Karasu Valley, Uzbekistan

Climate is regarded to be a driving force behind deposition, erosion, diagenesis and reworking of deposits, and, on the other hand, soil formation and vegetation development. Therefore, loess-soil successions can be viewed as effects of changing environmental conditions, which in turn are strongly correlated with climate fluctuations (Derbyshire et al. 1995; Kraus 1999; Jary 2007; Sheldon and Tabor 2009; Welc et al. 2016). During periods of relatively warm and wet climate, deposition of aeolian fine fraction was strongly limited, resulting in the development of pedogenic cover on surfaces stabilized by vegetation. Contrastingly, during dry and cool intervals, decrease in vegetation cover enhanced physical weathering, erosion, and transport of fine material that buried previously formed soils. Thus, pedocomplexes were formed, with relatively well developed paleosols (fossil soil) separated by large thicknesses of pedogenically-unmodified sediments. Usually, paleosols show significantly higher magnetic susceptibilities than do loess horizons, which aids with their recognition. Currently, a multi-proxy approach to paleosols allows both qualitative and quantitative paleoenvironmental and paleoclimatic reconstructions (Derbyshire et al. 1995; Kraus 1999; Jary 2007; Sheldon and Tabor 2009; Welc et al. 2016).

Fig. 2 **a** Holocene loess-soil exposure on the Karasu River bank, Tien Shan foothills, Uzbekistan. *Photo* M. Suska-Malawska (after Welc et al. 2016). **b** Northern shore of the Rangkul Lake, Eastern Pamir, Tajikistan. *Photo* M. Mętrak

At the foothills of the Tien Shan in the northern Uzbekistan, a unique series of the Holocene polygenetic soils interbedded with loess sequences can be found (Fig. 1b). They contain a full spectrum of short-term Holocene climatic oscillations in high resolution and as such, are exceptional on the global scale (Kovaleva 2004; Welc et al. 2016). However, preceding publications concerning Holocene and Late Pleistocene loess-soil sequences in Central Asia provide only a simplified scenario of climate changes, based mainly on $\delta^{13}C$ analysis (Kovaleva 2004, 2005). Hence, we decided to create a multi-proxy data base to complement the existing knowledge. Here we present preliminary results regarding magnetic susceptibility, as the studied soils are characterized by highly diversified values of this parameter that reflect changes in their formation processes.

Located in the central part of the Chirchik River-basin, on the left bank of the Karasu River, the Holocene loess-soil series in the Karasu Valley forms a vertical wall, several meters high, created by erosive activity of the Karasu River (Fig. 2a). The exposure reveals almost horizontal sequence of the loess deposits, alternating with paleosoil horizons, and alluvial silt and gravel sediments. Up to now, only a preliminary study of the Karasu profile was performed, covering data on paleomagnetism and magnetic susceptibility, based on 24 oriented samples from the exposure (Welc et al. 2016).

Lithological composition of the 5 m high Karasu Valley profile is as follows: (1) modern soil (SG-1), grey colour, about 0.30 m in thickness; (2) the silty loess of dark-grey colour with a thickness of 2.40 m; (3) upper paleosol horizon (PG—2), grey colour, about 0.25 m in thickness; (4) the silty loess of dark-grey colour with a thickness of 0.15 m; (5) lower paleosol horizon (PG—3), grey colour, about 0.50 m in thickness; (6) the silty loess of dark-grey colour with a thickness of 0.15 m; (7) layer of alluvial muds, of dark-grey colour, with interbeddings of gravels and pebbles, about 0.20 m in thickness; (8) layer of pebbles with diverse diameters (Welc et al. 2016). According to Professor Khodjiakbar Toychiew (personal communication), we can estimate the Holocene series in Karasu to be approximately 13,000 years old (on the basis of paleomagnetic scale). Current lack of radiocarbon dates prevents any further attempts to interpret the paleoclimatic aspects of the exposure. However, there is no doubt that the profile reveals a very detailed scenario of climate fluctuations, especially the short-term ones, as, for instance, the Bond cycles (Welc et al. 2016).

Lithological structure of the presented profile is typical for the second terrace of the Karasu River and reflects several sedimentation episodes that took place under different environmental conditions (Fig. 3). Loess layers were deposited during dry and windy periods, probably with occasional heavy rainfalls, while soils were formed during stable and relatively moist intervals. Bottom part of the profile is characterized by alluvial sediments of different granulometric composition, i.e. muds and the lowest lying river gravels. Interestingly, no correlation between paleosols occurrence and magnetic susceptibility were observed. Maximum values of magnetic susceptibility were recorded for material from the second horizon, comprising over 2 m of silty loess (Welc et al. 2016). This untypical situation was probably caused by complicated genesis of loess-soil successions in the Chirchik

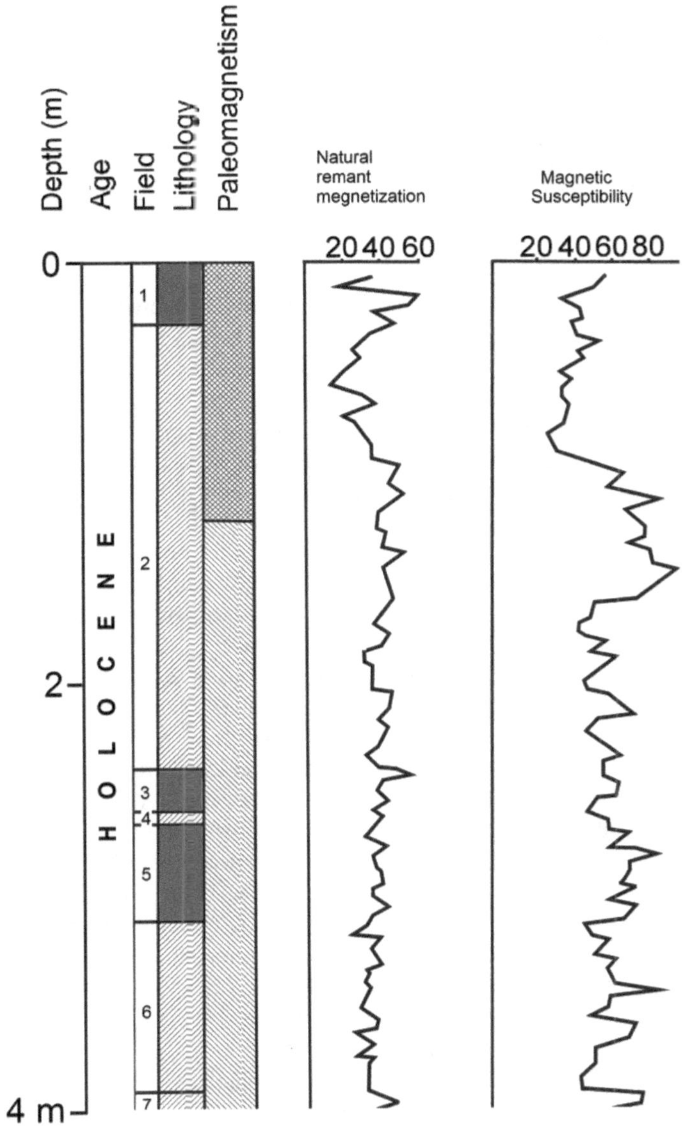

Fig. 3 Magnetic susceptibility and natural remnant magnetization recorded in Karasu loess-soil succession. Drawing by F. Welc, after Khodjiakbar Toychiew (Welc et al. 2016)

River basin. Large loess deposits in the north-east of Uzbekistan, including Karasu Valley deposits, seem to be formed according to a polygenetic formation theory, which favours a proluvial mechanism (in general, deposition by water on plains) (Yeliseyev 1973; Smalley et al. 2006; Welc et al. 2016). However, recent studies suggest, that deposits in the Chirchik River-basin could be additionally topped with typical aeolian deposits (Smalley et al. 2006).

3 The Glaciers: Development of High-Mountain Lakes on the Example of the Rangkul Lake, Eastern Pamir, Tajikistan

The Pamir Mountains, located on the orogenic uplift known as the Pamir Knot, joining several Asian mountain ranges, are characterized by extreme climatic conditions and rather short growing season. While the western part of Pamir is characterized by large relative elevations (Peak Somoni, 7495 m a.s.l, the highest summit of the Tajik Pamir is located here), narrow ravines with steep slopes and rapid mountainous rivers, the eastern part is an elevated plateau (3500–4000 m a.s.l.) with vast flat valleys with either meandering rivers or dry riverbeds (Shahgedanova 2002; Mętrak et al. 2015).

Though in general the climate of Pamir is arid continental, there are significant differences between western and eastern part of the mountains. While the Western Pamir receives around 1,500 mm of precipitation annually, with maxima reaching 2,500 mm for the Fedchenko Meteorological Station, in the Eastern Pamir, located in the rain shadow of high western peaks, mean annual precipitation ranges between 50 and 150 mm (Rojan 2007; Kayumov and Rajabov 2008; Aizen 2011; Vielicko and Spasskaa 2011; Mętrak et al. 2015). By contrast with the Western Pamir, rainfall in the east increases during summer, whereas during most winters high plateaus get no snow. Small precipitation rates, combined with high radiation rates, strong winds, and subzero average temperatures from October to March, make the Eastern Pamir a cold mountainous desert with strongly arid climate conditions (Rojan 2007; Kayumov and Rajabov 2008; Aizen 2011; Vielicko and Spasskaa 2011; Mętrak et al. 2015). These harsh conditions result in local and altitudinal environmental differences, especially as soil moisture, salinity and nutrient content are concerned, which in turn determine biological diversity of the Eastern Pamir. The main climate and environmental distinctions between the western and the eastern part of Pamir are summarized in Table 1.

Despite these desert characteristics, there are many high-mountain lakes in the Eastern Pamir, usually surrounded by wetlands, which play an important role as pastures in the otherwise barren, rocky mountain terrain, covered by moraine debris. These lake-wetland systems are supplied with waters from various sources, including scarce precipitation, surface and ground waters of glacial origin (glaciers, permafrost, glacial rivers, snowmelt) and geothermal springs, which tend to extend growing period in these ecosystems. As these varied sources influence development and functioning of Pamir lakes and wetlands, the ecosystems in question are especially vulnerable to climate changes, which are usually well recorded in the physical and chemical features of their bottom sediments (Mętrak et al. 2015).

Table 1 Environmental distinctions between the Western and the Eastern Pamir (Aizen 2011; Rojan 2007, after Mętrak et al. 2015)

	Western Pamir	Eastern Pamir
Relief	Low (ca. 1500 m a.sl.) and high (over 7500 m a.s.l.) ranges: **large relative elevations**	Elevated plateau (3000–4000 m a.s.l.) and high peaks (ca. 6000 m a.s.l.): **low relative elevations**
Valleys	Deep ravines with steep slopes and rapid rivers	Flat valleys with meandering rivers or dry riverbeds
Annual precipitation (mm)	1,500 (with maxima reaching 2,500 for the Fedchenko Meteorological Station)	50–150
Glaciation area (km^2)	6,110	1,454
Vascular plants	Over 1,500 species	Over 700 species

Hence, our aim was to reconstruct the history of chosen lakes and wetlands in the Eastern Pamir, in order to identify their response to paleoclimatic changes and to complement the existing records on the above mentioned alterations.

To assess the high mountain lakes' response to past climatic changes, we chose the Rangkul Lake, located in the eastern part of the Pamir Mountains at the elevation of about 3700 m a.s.l. Currently, the Rangkul Lake covers the area of around 8 km^2 with the maximum depth reaching some 1–2 m. Water supply to the Rangkul Lake is limited due to its location on the wide and flat platform. This platform formerly was a lake bottom (littoral part), yet due to the water level changes it become dry land, partially covered by an extensive fen meadow situated in the eastern part of the platform. Nowadays, this flat area, locally with vegetation, acts as a barrier stopping finer and coarser sediments, coming from the erosion of the mountain slopes. General characteristics of the Rangkul Lake are presented in Table 2.

Table 2 General characteristics of the chosen lakes in the Pamir (after Mętrak et al. 2015)

Lake	Altitude (m a.s.l.)	Area (km^2)	pH	Summer temperature (°C)	EC (mS/cm)	SAL (mg/l)	DO (mg/l)	Author
Rangkul	3784	7.8	6.4–9.2	16.5	x	355–463	x	Achrorov (2006)
	x	8.3*	7.81–9.54	11.9–12.1	0.07–1.3	30–650	5.85–6.03	Mętrak et al. (2015)

EC electric conductivity; *SAL* salinity; *DO* dissolved oxygen
*Own unpublished data

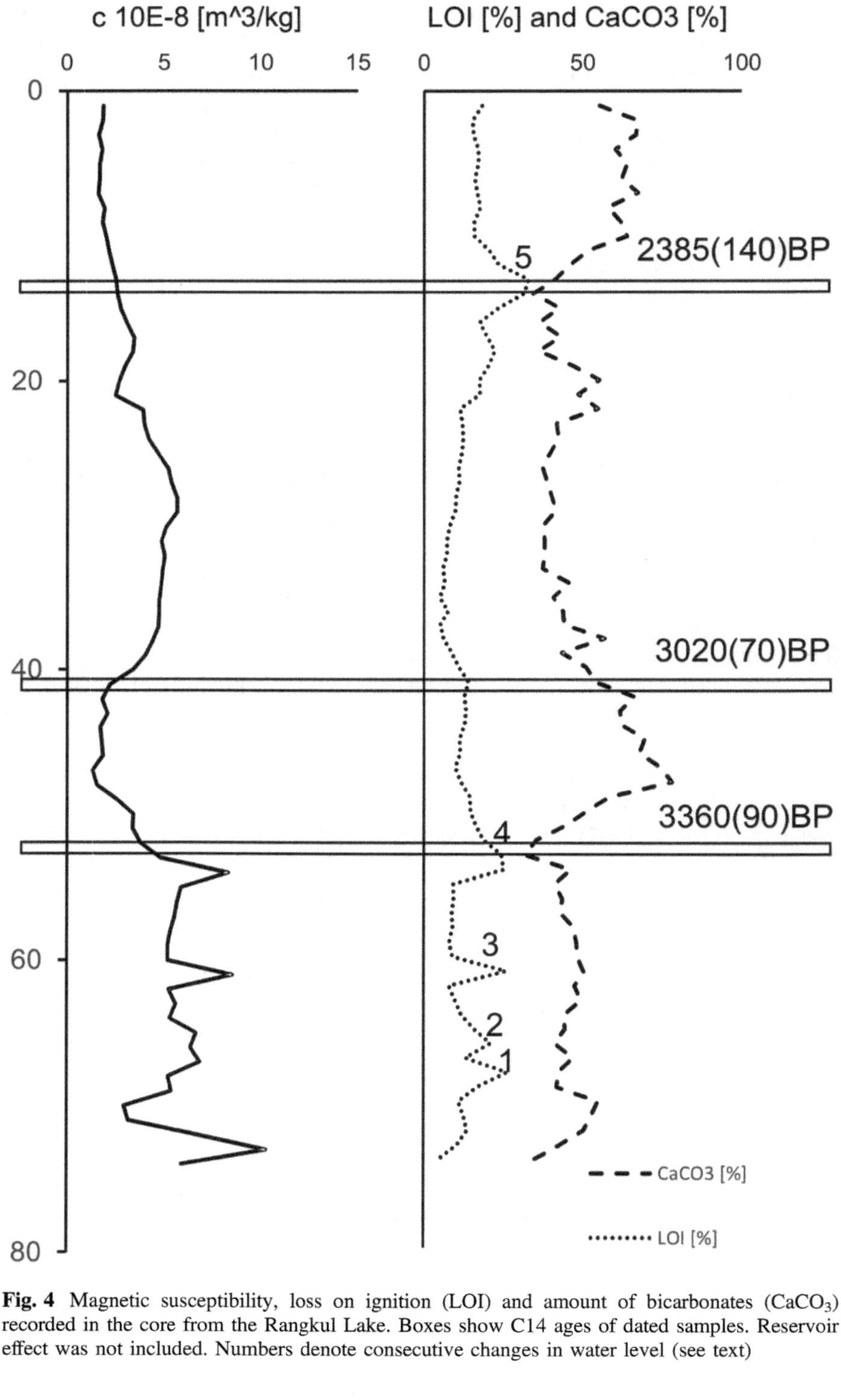

Fig. 4 Magnetic susceptibility, loss on ignition (LOI) and amount of bicarbonates ($CaCO_3$) recorded in the core from the Rangkul Lake. Boxes show C14 ages of dated samples. Reservoir effect was not included. Numbers denote consecutive changes in water level (see text)

Out of the Rangkul Lake we took a core for multi proxy analyses using the UWITEC auger (0.74 m long, divided into samples of 1 cm each, giving a total of 74 samples). Magnetic susceptibility, loss on ignition and amount of bicarbonates were measured in every sample. According to the obtained results, we chose samples for further analyses, i.e., SEM analyses, elemental analyses and C14 dating. Selected results are presented in Fig. 4.

Rangkul is a relatively young lake, characterized by rather low sedimentation rate of the surface horizons. As the age is concerned, we must take into consideration a reservoir effect, that in the Pamir lakes is estimated at 800 years (Mischke et al. 2010). Up to now, no detailed results regarding the Rangkul Lake are available, as we are waiting for C14 analyses of biomass samples from this lake. Nevertheless, we expect this lake to be around 800 years younger than the basic datings presented in this paper. During development of the Rangkul Lake 5 events of significant decrease in water level occurred, which caused major changes in type of sediments. Over the periods of low water level, wetland vegetation developed on partially uncovered lake bottom, resulting in formation of organic sediments. These shifts are clearly visible in Fig. 4 as peaks in loss on ignition values, signed with consecutive numbers. Four episodes of drying occurred before 3360 BP, while the last one happened around 2385 years ago (no reservoir effect included).

Though usually a fast sedimentation of weakly magnetic or non-magnetic biogenic sediments causes dilution of highly magnetic terrigenous materials (Geiss et al. 2003), in case of events number 2, 3 and 4 organic matter influxes coincided with an increase of magnetic susceptibility. Over dry periods in the areas distant from the lake (including mountain slopes), the vegetation cover become sparse or disappeared totally, hence the intensification of erosion processes, that in turn increased the supply of strongly transformed, highly magnetic material to the lake. Moreover, magnetic minerals (e.g. magnetite) could have been produced in microbial processes (Hu et al. 2002), that boosted in well-aerated and still moist wetland habitats around the shrunken Rangkul Lake. However, biogenic processes are much slower in comparison to terrigenous sediment deposition.

Over periods of relatively high water level, wetlands around the Rangkul Lake diminished and mineral fraction in lake sediments reached again over 90%, with huge participation of bicarbonates (more than 50%), especially in periods between 3360 and 3020 BP and after 2385 BP (no reservoir effect included). With the increase in bicarbonates, decrease in magnetic susceptibility can be observed, which is in accordance with accessible literature (Feng and Johnson 1995; de Jong et al. 2000; Issmer 2010).

After performing further analyses, more detailed interpretation of the Rangkul Lake development can be prepared. While interpreting obtained data, we must keep in mind that some of the textural and geochemical changes in sediments of the Rangkul Lake, especially in its top parts, might have been caused by human economic activities, both in prehistoric and modern periods (e.g. animal breeding).

4 Summary

Though magnetic parameters of rocks and sediments are not paleoclimate proxies per se (Geiss et al. 2003), they can be interpreted in terms of climate after establishing models linking magnetic variation to climatic change. In loess, where parent material is constant, magnetism acts as proxy for annual precipitation and to a lesser extent for annual temperature (Geiss et al. 2003). In case of lake sediments, situation is much more complicated; hence, the magnetic response of lacustrine sediments to climatic change may vary widely from lake to lake (Hu et al. 2002). However, having in mind all restrictions of this method, magnetic properties are a useful part of multiproxy paleoenvironmental and paleoclimatic analyses.

Acknowledgements Authors wish to thank Professor Khodjiakbar Toychiew for his scientific support and consultations.

Funding This work was supported by the National Science Centre (Grant No 2013/09/B/ST10/01662).

References

Achrorov F (2006) Hydrography and productivity of high-mountain lakes in the Pamir. Academy of Sciences of Tajikistan. Works vol. 1 (27), Dushanbe

Aizen VB (2011) Pamir glaciers. In: Singh VP, Singh P (eds) Encyclopedia of snow, ice and glaciers. Haritashya Springer Publisher, U.K, p 1253

Baibatsha AB (2012) Anthropogenic history of Kazakhstan. Almaty

Dambricourt MA, Gaillard C (2011) Relations between climatic changes and prehistoric human migrations during Holocene between Gissar Range, Pamir, Hindu Kush and Kashmir: the archeological and ecological data. Quatern. Int. 229:123–131

Derbyshire E, Kemp R, Meng XM (1995) Variations in loess and paleosol properties as indicators of paleoclimatic gradients across the Loess Plateau of north China. Quatern. Sci. Rev. 14: 681–697

de Jong E, Pennock DJ, Nestor PA (2000) Magnetic susceptibility of soils in different slope positions in Saskatchewan, Canada. CATENA 40:291–305

Encyclopædia Britannica Online, s. v. "Central Asia". Accessed 06.05. 2016, http://www.britannica.com/place/Central-Asia

Feng ZD, Johnson WC (1995) Factors affecting the magnetic susceptibility of a loess-soil sequence, Barton County, Kansas, USA. CATENA 24:25–37

freeworldmap.net [access 10.05.2016]

Geiss CE, Umbanhowar CE, Camill P, Banerjee SK (2003) Sediment magnetic properties reveal Holocene climate change along the Minnesota prairie-forest ecotone. J. Paleoclim. 30:151–166

Glantz MM, Viola B, Chikisheva T (2004) New hominid remains from Obi-Rakhmat Grotto. In: Derevianko AP (ed) Grot Obi-Rakhmat. Institute of Archeology and Ethnography, Siberian Branch, Russian Academy of Sciences, Novosibirsk, pp 77–93

Hu S, Deng C, Appel E, Verosub KL (2002) Environmental magnetic studies of lacustrine sediments. Chin. Sci. Bull. 47:613–616

Islamov UI, Zubov AA, Kharitonov VM (1988) Paleoliticheskaya stoyanka Sel Ungur v Ferganskoi doline. Voprosy Antropologii 80:38–49

Issmer K (2010) Wpływ cech litologicznych na podatność magnetyczną lessów z rejonu Wzgórz Dalkowskich, Badania fizjograficzne, Seria A, Geografia fizyczna (A61), s. 169–180. Doi:10. 2478/v10116-010-0010-4

Jary Z (2007) Zapis zmian klimatu w górnoplejstoceńskich sekwencjach lessowo-glebowych w Polsce i w zachodniej części Ukrainy. Rozprawy Naukowe Instytutu Geografii i Rozwoju Regionalnego Uniwersytetu Wrocławskiego, Wrocław

Kayumov A, Rajabov I (2008) Glacier – water resources in Tajikistan in condition of climate change. Dushanbe; [at www.meteo.tj. Accessed on 09.10.2014]

Kovaleva N (2004) Northern Tian Shan paleosoil sedimentary sequences as a record of major climatic events in the last 30,000 years. Revista Mexicana Ciencias Geológicas 21:71–78

Kovaleva N (2005) Paleoclimatic significance of the loess-paleosol sequences from Tien Shan, Central Asia. Geophysical Research Abstracts 7. SRef-ID:1607-7962/gra/EGU05-A-01339

Kraus MJ (1999) Paleosols in clastic sedimentary rocks: their geologic applications. Earth Sci. Rev. 47:41–70

Mętrak M, Sulwiński M, Chachulski Ł, Wilk M, Suska-Malawska M (2015) Creeping Environmental Problems in the Pamir Mountains: Landscape Conditions, Climate Change, wise Use and Threats. In: Öztürk M, Hakeem KR, Faridah-Hanum I, Efe R (eds.) Climate change impacts on high-altitude ecosystems. Springer. doi:10.1007/978-3-319-12859-7

Mischke S, Rajabov I, Mustaeva N, Zhang C, Herzschuh U, Boomer I, Brown ET, Andersen N, Myrbo A, Ito E, Schudack ME (2010) Modern hydrology and late Holocene history of Lake Karakul, eastern Pamirs (Tajikistan): A reconnaissance study. Paleogeo. Paleoclim. Paleoecol. 289:10–24

Ranov V, Davis R (1979) Toward a new understanding of the Soviet Central Asian Paleolithic. Curr. Anthropol. 20:249–270

Rojan E (2007) The morphogenetic role of glaciers in the north-western Pamirs. Słupckie Prace Geograficzne 4:123–133

Shahgedanova M (ed) (2002) Physical geography in northern Eurasia. Oxford University Press, Oxford, pp 1–592

Sheldon ND, Tabor NJ (2009) Quantitative paleoenvironmental and paleoclimatic reconstruction using paleosols. Earth Sci. Rev. 95:1–52

Smalley IJ, Mavlyanova NG, Rakhmatullaev KhL, Shermatov MSh, Machalett B, O'Hara Dhand K, Jefferson IF (2006) The formation of loess deposits in the Tashkent region and parts of Central Asia; and problems with irrigation, hydrocollapse and soil erosion. Quatern. Int. 152–153:59–69

Wang X, Sun D, Chen F, Wang F, Li B, Popov SV, Wu S, Zhang Y, Li Z (2014) Cenozoic paleo-environmental evolution of the Pamir-Tien Shan convergence zone. J. Asian Earth Sci. 80:84–100

Welc F, Toychiew K, Suska-Malawska M, Marks L, Mętrak M (2016) Paleoclimatological and geoarcheological significance of the Holocene loess-soil successions of the Tien Shan foothills of Uzbekistan. Studia Quaternaria 33:57–68

Vielicko AA, Spasskaa II (2011) Approaches to assessment of relief-forming processes under conditions of global warming (with reference to Northern Eurasia within the boundaries of the former USSR). Geographia Polonica 84(Special Issue, Part 1):179–187

Yeliseyev VI (1973) On the loess soils of Middle Asia and Kazakhstan. Byulleten Komissii po Izucheniya Chetvertinchnogo Perioda 40:52–68 (in Russian)

Zhang Z, Han W, Fang X, Song Ch, Li X (2013) Late Miocene-Pleistocene aridification of Asian Inland revealed by geochemical records of lacustrine-fan delta sediments from the western Tarim Basin, NW China. Palaeogeogr. Palaeoclimatol.0 Palaeoecol. 377:52–61

Application of Magnetic Susceptibility Measurements for Identification of Technogenic Horizons in Soil Profiles on the Example of the Vistula River Cross-Cut Area

Grzegorz Kusza, Piotr Hulisz, Leszek Łęczyński, Adam Michalski, Michał Dąbrowski and Żaneta Kłostowska

Abstract Studies on magnetic properties of soils have been recently applied in the soil environment quality monitoring in relation to basic physical and chemical properties. However, the issue of distribution of ferrimagnetic materials in particular genetic soil horizons, especially in altered soils, has not yet been considered in environmental studies. The main subject of the research was to evaluate a potential of magnetic susceptibility measurements as implements for supporting the soil classification through indication of particular distinctive genetic horizons in the soil profile. The study objects were soils whose formation was strictly conditioned by hydrotechnical regulations in the area of the Vistula River Cross-Cut (northern Poland). It has been demonstrated that the magnetic susceptibility test is an accurate supporting tool in recognition and classification of genetic horizons in a soil profile. The obtained results properly reflected the soil morphological variability and also were good indicators of the presence of lithologic discontinuities of technogenic origin.

Keywords Magnetic susceptibility · Technogenic horizons · Vistula river Cross-Cut

---

G. Kusza (✉)
Department of Land Protection, University of Opole, Oleska 22,
45-052 Opole, Poland
e-mail: Grzegorz.Kusza@uni.opole.pl

P. Hulisz · A. Michalski · M. Dąbrowski
Faculty of Earth Sciences, Department of Soil Science and Landscape Management,
Nicolaus Copernicus University in Toruń, Lwowska 1, 87-100 Toruń, Poland

L. Łęczyński · Ż. Kłostowska
Institute of Oceanography, Laboratory of Applied Geology, University of Gdańsk,
Piłsudskiego 46, 81-378 Gdynia, Poland

M. Jeleńska et al. (eds.), *Magnetometry in Environmental Sciences*,
GeoPlanet: Earth and Planetary Sciences, DOI 10.1007/978-3-319-60213-4_5

1 Introduction

Hydrotechnical rebuilding of river-beds is one of the factors that affect the changes of the natural environment of river terraces, particularly soil conditions. It changes both, geological conditions as well as natural processes of soil formation.

The Vistula River Cross-Cut, made in 1895 to provide flood protection of the Żuławy Wiślane area, is the most radical and influential hydrographic change in the river mouth area. The river engineering has caused a strong transformation of the natural environment, of which the main result was the formation of a new mouth of the Vistula River (Makowski 1995; Robakiewicz 2010). Material excavated during the digging works has been lodged directly at the new river bed, forming embankments. Consequently, specific soils have been formed with clearly visible anthropogenic horizons flat-lying on natural genetic ones in the initial sedentary soil. Another element of hydrotechnical works affecting the formation of the soil cover was the construction of breakwaters, which on the one hand were supposed to protect the coastline against ice-jams, and on the other hand, they were designed to enable navigation by reducing marine and fluvial sedimentation (delta cones and sandbars with shallow lakes) (Wróblewski et al. 2015). Moreover, natural factors related to the land use structure constantly affected the soil cover in the area adjacent to the Cross-Cut. Almost the whole area has been forested with artificial planting of pine and small fragments of deciduous forest. The local forest management focuses on the cultivation of coniferous trees which results in pH lowering and leaching of base cations from the surface horizons of soils. Considering the reserve of "Mewia Łacha" near the estuary cone, special protection area for birds in Natura 2000 "Mouth of the Vistula" PLB220004 (no vegetation cover), together with the adjacent dunes, the area is potentially threatened by displacement of fine sand fraction in the process of aeolian erosion and subsequent deposition on adjacent areas. Changes in the genetic horizons arrangement due to dynamic processes, including degradation, caused the need for research which aimed at identifying layers of the technogenic origin and evaluation of the potential level of soil degradation. One of the methods currently used in the assessment of changes in the environment is the measurement of magnetic parameters (Thompson and Oldfield 1986; Oldfield 1991; Strzyszcz et al. 1994; Dearing et al. 1996; Gelisli and Aydin 1998; Han and Zhang 2013). Magnetic properties are used to assess soil pollution, both in terms of deposition of industrial pollutants, including heavy metals (Strzyszcz 1993; Strzyszcz and Magiera 1998; Kapicka et al. 2001; Hanesch and Schlogfer 2002; Magiera et al. 2007; Duan et al. 2010), as well as changes caused by natural factors (Magiera et al. 2006; Hanesch et al. 2007). The measurement of magnetic susceptibility is a very fast and precise method for the analysis of changes induced by anthropogenic factors (Strzyszcz 1993; Dearing 1999). Magnetic measurements are also used in the analysis of the spatial distribution of pollutants of technogenic origin (Strzyszcz and Rachwał 2008; Zawadzki et al. 2009, 2010).

The aim of this study was to evaluate the applicability of magnetic susceptibility measurements for identifying the genetic horizons of soils both transformed by human pressure (soils of embankments formed during hydrotechnical works), as well as created through the natural sedimentation processes in the region of the Vistula River Cross-Cut on the background of selected physical and physico-chemical soil properties.

2 Materials and Methods

2.1 Study Area

The study was conducted within the area of piercing—the new river channel of the Vistula River in its outlet (Fig. 1).

Soil cover of the study area consists mainly of soils formed from sands of marine, fluvial, aeolian and technogenic origin (initial soils and Arenosols in various stages of development). Spatial pattern of the soil cover can be described as three soil belts interrupted by semihydrogenic (Eutric Fluvic and Eutric/or Fluvic Gleysols, Fig. 1) and technogenic soils (Eutric Regosols and Albic Arenosols, Fig. 1). The first belt is a coastal area dominated by beach sands and soils in initial stage of development. The second one, lying southward, consists mainly of Eutric and Albic Arenosols developed also from marine sands—see profile 1 (Fig. 1). The third belt is represented by Albic Arenosols formed from aeolian sands of former Vistula Spit. Soils within these belts were formed by natural sedimentation processes altered by human activity (built of breakwaters stimulating and accelerating sedimentation, forest management). Quite different soils developed within artificial embankments, which were formed as a result of accumulation of deposits of Vistula River Cross-Cut during hydrotechnical works. Soils of these embankments are represented by Eutric Regosols (Relocatic, Humic) or Albic Arenosols (Relocatic) —see profiles 2 and 3, Fig. 2. Despite technogenic origin they cannot be classified as Technosols, in the meaning of WRB classification system, because of the lack of artefacts in the upper part of soil profiles. Their technogenic genesis is emphasized by supplementary qualifier Relocatic. It indicates process of in situ remodelling of soil profile by human activity to a depth $\geq$ 100 cm and lack of horizons other than epipedon A.

Origin of many soil profiles is complicated because of alternate influence of fluvial, marine and aeolian processes altered by human activity. High dynamics of these processes resulted in formation of multilayered profiles with burial horizons or pedons (see profile 3, Fig. 2).

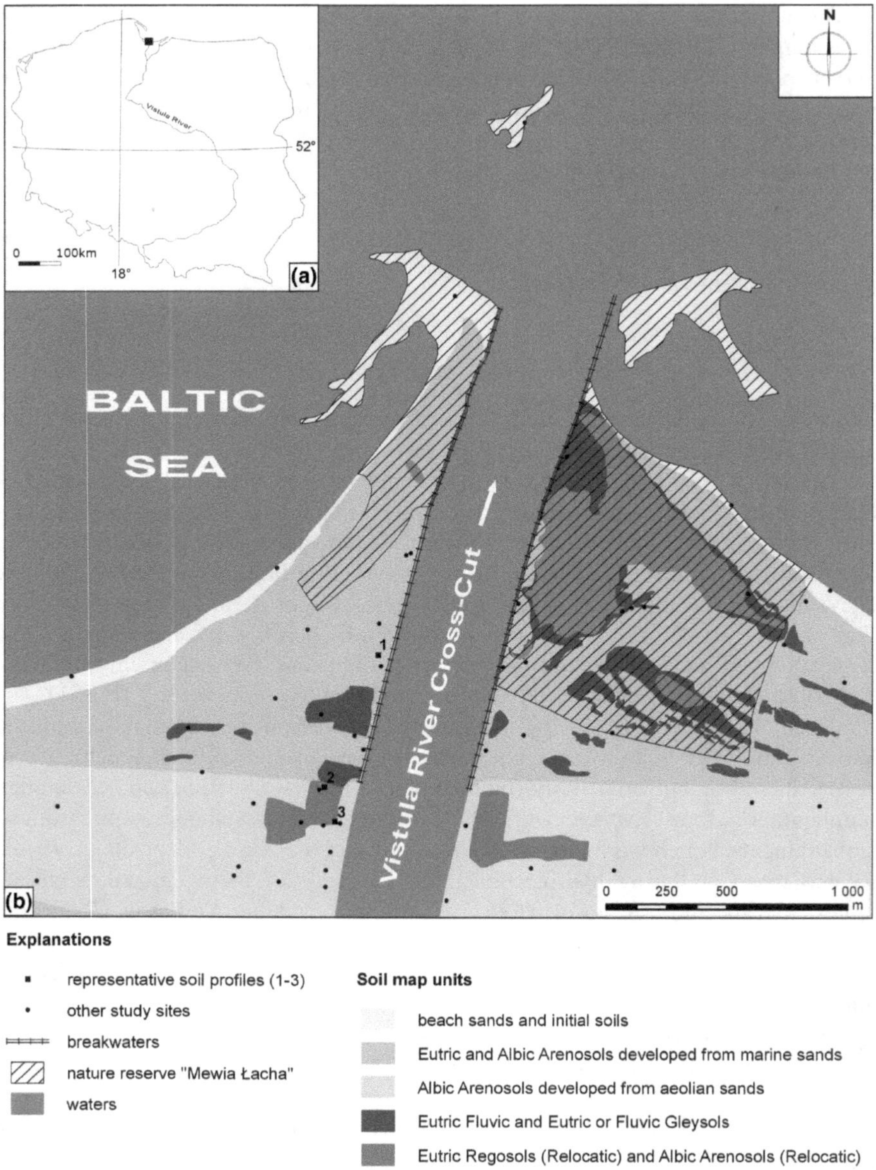

Fig. 1 Location (**a**) and soil map of the study area (**b**) Hulisz et al. (2015), modified

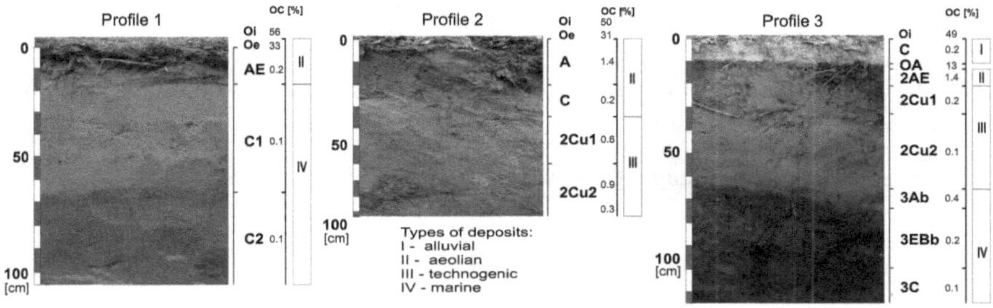

Fig. 2 Examples of soil profiles located on the left bank of the Vistula River Cross-Cut (Hulisz et al. 2015, modified): *profile 1* Albic Arenosol, *profile 2* Eutric Regosol (Relocatic, Humic), *profile 3* Albic Arenosol (Relocatic) over Podzol. Symbols of soil horizons according to Guidelines for Soil Description (2006)

3 Methods

Identification of soils occurring in the area of the Vistula River Cross-Cut outlet and the analysis of their spatial distribution were carried out directly during the field work. Soil pits were done on the most representative areas selected for each soil map unit (Fig. 1; Hulisz et al. 2015). Samples of disturbed soil structure were collected from each morphologically distinctive genetic horizon. Then samples were transported to the laboratory and dried in a dark place at room temperature. The homogenized samples were sifted through a sieve of 2 mm mesh size to separate skeletal parts. The sieved material was analyzed according to standard methods used in soil science.

The following soil properties were determined:

- particle size distribution by the sieve method and the hydrometer method (the Bouyoucos areometric, modified by Cassagrande and Prószyński), according to USDA,
- the pH value of the soil-to-solution ratio of 1:2.5 using H_2O and 1 M KCl as the suspension medium,
- electrical conductivity (EC) of soil-water extract 1:5 by conductometric method,
- content of carbonates by the Scheibler volumetric method,
- the total organic carbon (TOC) and total nitrogen (N_t) analysed on Vario MaxCube CN Elementar analyser in steel cylinders,
- LOI analyze by the samples ignition at temperatures of 550° in a muffle furnace.

3.1 Magnetic Parameters

The specific (mass) magnetic susceptibility (χ) was determined with the use of the MS2 "Bartington" laboratory magnetic susceptibility meter, equipped with a dual frequency MS2B sensor—low frequency (κlf) is 0.47 kHz and high frequency (κhf) is 4.7 kHz.

The following magnetic parameters were calculated in accordance with the appropriate equation:

(1) The specific (mass) magnetic susceptibility (χ):

$$\chi = \frac{\kappa}{\rho}$$

where:

κ low-field magnetic susceptibility, measured by "Bartington" susceptibility meter
ρ the density of a soil placed in standardised Bartington 10 cc plastic container and weighed ($m^3\ kg^{-1}$)

The magnetic susceptibility (κ) is expressed in dimensionless SI units. Values of specific (mass) magnetic susceptibility (χ) are given in $m^3\ kg^{-1}$ (Thompson and Oldfield 1986).

4 Results and Discussion

The investigated soil profiles were characterized by a significant diversity of genetic horizons, which were derived from both natural processes (marine and fluvial sedimentation and aeolian erosion), as well as human activity (soil layers from the material obtained during hydrotechnical works). Soils selected for the study reflect the dynamics of processes occurring in the area.

Soils represented by the profile 1 (Albic Arenosols) were developed from the marine sands but currently the impact of seawater is minimal. Deep and intense sorting of sand fraction has been evident. Sand fraction of 0.25–0.5 mm dominated in the grain size composition with percentage in the surface horizon AE (0–20 cm) of 75% and over 80% in the parent rock C1 and C2. There was no contribution of sand fraction of less than 0.1 mm revealed. Such a texture directly favours the very low capacity of sorption complex and supports very fast processes of nutrient and trace element leaching downwards the soil profile. Low sorption ability of the soil is also an effect of very low contents of the organic matter. Negligible content of organic carbon of 0.24% was recorded just below the organic horizon. This has been confirmed by the results of LOI analysis wherein mineral layers showed practically no organic matter content. Results of the reaction analysis indicated a decrease in pH values in the upper Oi, Oe and AE horizons which should be linked to acidic litter arising from the fallout of pine *Pinus sylvestris* needles. In the deeper layers of the profile, an increase of pH values (in KCl) from 4.1 to 5.6 was recorded, which indicated the influence of groundwater infiltrating deep into the analyzed

Table 1 Selected chemical properties of soils

Genetic horizon	Depth (cm)	EC (μS cm^{-1})	LOI (%)	TOC (%)	Nt (%)	C/N	pH	
							H_2O	KCl
Profile 1								
Oi	3–1	21.1	97.2	55.5	0.545	102	4.0	3.7
Oe	1–0	26.4	54.9	32.8	1.061	31	4.6	4.1
AE	0–20	12.9	0.53	0.24	0.011	23	4.4	3.8
C1	20–30	15.6	0.21	–	–	–	4.7	4.1
C1	30–40	14.1	0.13	–	–	–	5.2	4.6
C1	40–55	14.2	0.19	–	–	–	5.5	4.7
C1	55–65	18.0	0.11	–	–	–	5.7	5.2
C2	65–75	14.7	0.13	–	–	–	6.0	5.6
C2	75–110	13.4	0.11	–	–	–	5.9	5.5
Profile 2								
Oi	2–1	28.5	94.5	49.7	1.104	45	4.3	4.0
Oe	1–0	34.1	55.8	30.6	1.166	26	4.9	4.6
A	0–20	21.0	2.79	1.40	0.102	14	4.8	4.2
C	20–35	17.6	0.41	–	–	–	5.3	4.4
$2Cu_1$	35–60	23.9	2.15	0.60	0.052	12	4.6	3.5
$2Cu_2$	60–80	21.0	0.98	0.25	0.022	11	5.8	4.8
$2Cu_2$	80–90	30.4	0.92	0.92	0.081	11	5.3	4.3
Profile 3								
Oi	1–0	32.5	89.2	49.0	1.288	38	4.6	4.4
C	0–10	18.5	0.41	–	–	–	5.3	4.4
OA	10–12	34.7	22.5	13.0	0.619	21	3.9	3.1
2AE	12–22	20.3	2.73	1.38	0.079	18	4.0	3.2
$2Cu_1$	22–32	19.0	0.56	0.21	0.013	16	4.4	3.7
$2Cu_2$	32–65	18.6	0.15	–	–	–	4.8	4.4
3Ab	65–75	20.7	0.81	0.38	0.016	24	4.6	4.1
3EBb	75–100	15.0	0.34	–	–	–	4.7	4.1
3C	100–115	23.5	0.19	–	–	–	4.5	4.2

profile. A significant impact of grain size composition on soil processes has been revealed by values of electrical conductivity (EC). EC values recorded in the mineral horizons of the profile 1 ranged from 12.9 to 18.0 μS cm^{-1}, mirroring the soil texture and organic matter content (Table 1). The obtained values of low magnetic susceptibility (κ) did not exceed the value of 8.4×10^{-5} SI, confirming the lack of magnetic strengthening and thus the lack of ferromagnets of natural or technogenic origin. Similarly low values were obtained for the specific magnetic susceptibility (χ), with a maximum of 5.9×10^{-8} m^3 kg^{-1} recorded in the parent rock C2 at a depth of 65–75 cm (Table 2).

Table 2 Selected magnetic parameter values of soils

Genetic horizon	Depth (cm)	κlf ($\times 10^{-5}$ SI)	κhf ($\times 10^{-5}$ SI)	χ ($\times 10^{-8}$ m^3 kg^{-1})
Profile 1				
Oi	3–1	2.1	1.8	2.0
Oe	1–0	3.8	3.3	3.1
AE	0–20	4.5	4.0	4.0
C1	20–30	3.3	3.0	2.0
C1	30–40	1.3	1.1	0.8
C1	40–55	2.6	2.2	1.7
C1	55–65	2.3	2.0	1.3
C2	65–75	8.4	8.2	2.9
C2	75–110	7.3	7.1	4.8
Profile 2				
Oi	2–1	2.3	2.0	1.2
Oe	1–0	3.9	3.4	2.9
A	0–20	3.3	3.0	3.7
C	20–35	1.2	0.8	0.6
$2Cu_1$	35–60	6.0	5.8	4.3
$2Cu_2$	60–80	2.0	1.7	1.3
$2Cu_2$	80–90	8.0	7.7	7.3
Profile 3				
Oi	1–0	0.33	0.29	0.2
C	0–10	9.1	8.7	12.2
OA	10–12	5.67	5.4	3.9
2AE	12–22	3.33	2.6	2.1
$2Cu_1$	22–32	0.33	0.29	0.2
$2Cu_2$	32–65	4.1	3.6	2.5
3Ab	65–75	0.6	0.4	0.2
3EBb	75–100	0.0	0.0	0.0
3C	100–115	0.0	0.0	0.0

Analyzing (κ) values at different genetic horizons, a clear boundary between the horizons of C1 and C2 has been noted, as the (κ) MS values were twice higher in C2 level and amounted to 8.4×10^{-5} SI and 7.3×10^{-5} SI, respectively, at a depth of 65–75 cm and 75–110 cm. The obtained values pointed to the presence of horizons of different origin, e.g. from fluvial and marine deposits (Hulisz 2013; Zan et al. 2015).

The analysis of soil characteristics in profile 2, classified as Eutric Regosol (Relocatic, Humic), revealed the presence of two sorts of mineral deposits in the surface layers, which were built by sands of 0.5–0.25 and 0.25–0.1 mm (approx. 90%) from wind erosion (aeolian sands) and by finer technogenic deposits, mostly fine sandy loam and loamy sand (Table 3). Such a differentiation, in terms of the texture of particular horizons, has affected the properties of the individual, newly

Table 3 Particle size distribution

Genetic horizon	Depth (cm)	Percentage share of fraction (mm)										Textural class (USDA)
		>2.0	2.0–1.0	1.0–0.5	0.5–0.25	0.25–0.1	0.1–0.05	0.05–0.02	0.02–0.005	0.005–0.002	<0.002	
Profile 1												
AE	0–20	0	0	13	75	12	0	0	0	0	0	Fine sand
C1	20–30	0	0	5	83	12	0	0	0	0	0	Sand
C1	30–40	0	0	4	85	11	0	0	0	0	0	Sand
C1	40–55	0	0	7	83	10	0	0	0	0	0	Sand
C1	55–65	0	0	7	81	12	0	0	0	0	0	Sand
C2	65–75	0	0	15	78	7	0	0	0	0	0	Sand
C2	75–110	0	0	12	79	9	0	0	0	0	0	Sand
Profile 2												
A	0–20	0	0	3	37	52	3	3	1	0	1	Fine sand
C	20–35	0	0	6	40	50	2	1	0	0	1	Fine sand
$2Cu_1$	35–60	0	0	2	16	29	16	16	6	3	12	Fine sandy loam
$2Cu_2$	60–90	0	1	9	40	23	9	7	3	1	7	Loamy sand
$2Cu_2$	90	0	0	1	0	0	8	30	26	12	23	Silt loam
Profile 3												
C	0–10	0	8	50	42	0	0	0	0	0	0	Sand
2AE	12–22	0	0	2	38	55	3	0	0	2	0	Fine sand
$2Cu_1$	22–32	0	3	44	52	1	0	0	0	0	0	Sand
$2Cu_2$	32–65	0	1	32	67	0	0	0	0	0	0	Sand
3Ab	65–75	0	0	2	50	47	0	0	0	1	0	Sand
3EBb	75–100	0	1	54	45	0	0	0	0	0	0	Sand
3C	100–115	0	1	71	28	0	0	0	0	0	0	Sand

formed genetic horizons in heavily transformed soils. Only slight variations in the values of pH and EC were observed between the layer of aeolian sand and material deposited during civil engineering, indicating stabilization of soil forming processes. However, a significant change has been revealed in the measurement of both low magnetic susceptibility (κ) and the specific magnetic susceptibility (χ). Values of low magnetic susceptibility (κ) varied in each of the analyzed genetic horizons, with a maximum in the parent rock $2Cu_2$ at a depth of 80–90 cm. This expresses the movement of the ferromagnets particles from the surface layers and their gradual bonding in the deeper layers of the soil profile by fine fractions with a diameter of <0.05 mm.

In the case of the profile 3—Albic Arenosol (Relocatic) over Podzol—the occurrence of sandy deposits in the whole soil profile was observed. Some variation was revealed only in the fraction of 0.25–0.1 mm whose dominant share was characteristic for upper horizons, both in horizon 3 Ab of fossil soil resulting from marine sedimentation and in AE typical for the formation of soil by the process of aeolian erosion. Differentiation in EC values in all analyzed genetic horizons exhibited the effect of overlapping of different processes during soil evolution. A similar trend was observed in magnetic properties. Low magnetic susceptibility values showed only small magnetic amplification in the soil formed by marine sediments (being built of sand with a diameter of 1.0–0.5 and 0.5–0.25 mm), which had very low sorption properties. Slightly higher values were obtained in layers shaped by alluvial processes at a depth of 0–10 cm below the surface, where (κ) was 9.1×10^{-5} SI, as well as in the horizon resulting from the hydrotechnical

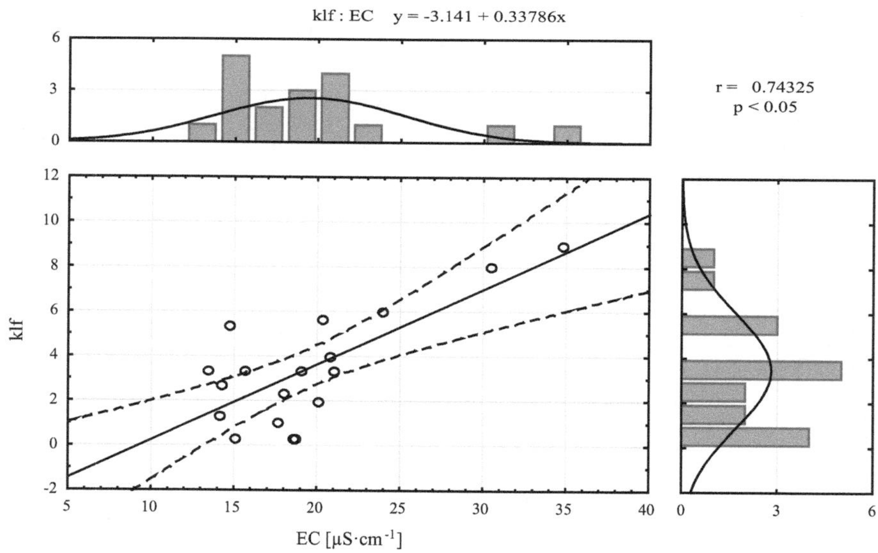

Fig. 3 Relationship between low magnetic susceptibility (κlf) and electrical conductivity (EC) of soils

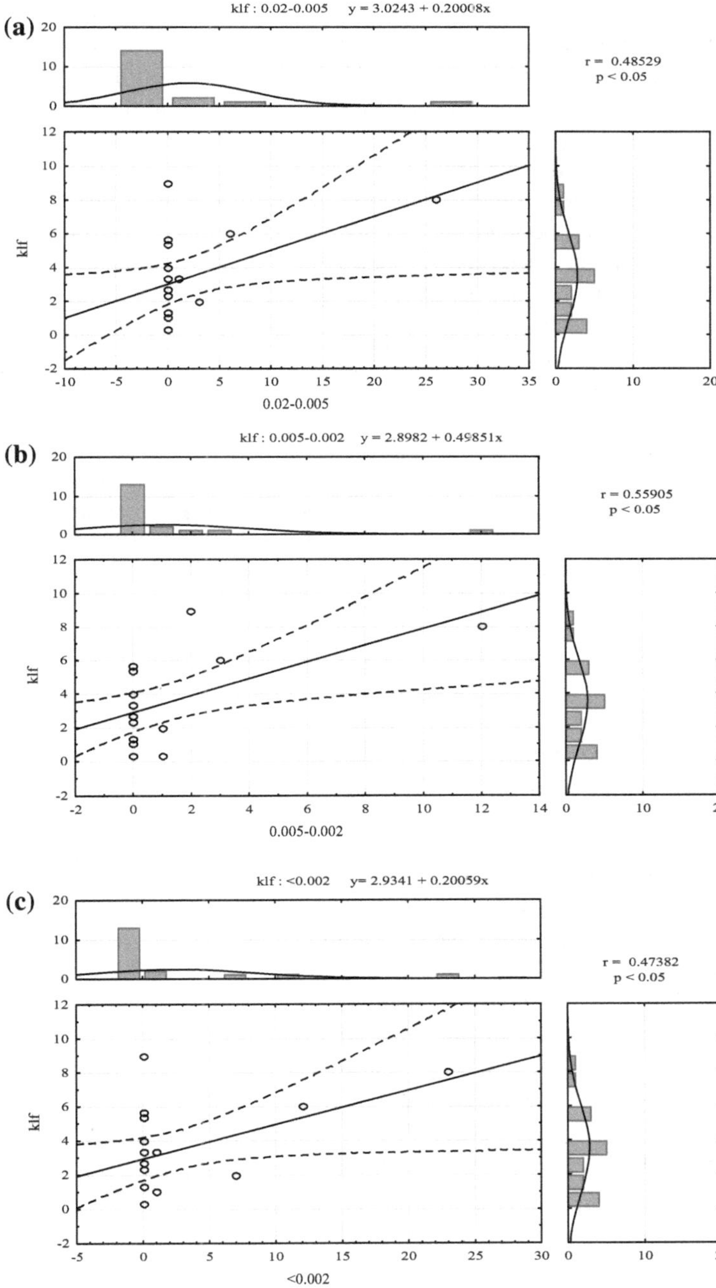

Fig. 4 Pearson's correlation coefficients between low magnetic susceptibility and fraction's share: **a** 0.02–0.005, **b** 0.005–0.002 and **c** <0.002

works during the construction of the Vistula River Cross-Cut. In the artificially deposited layers, the obtained values of (κ) ranged from 0.3 to 5.6 × 10^{-5} SI.

Considering the content of calcium carbonate, the results of the analysis showed the lack of the compound in all analyzed genetic horizons in the investigated soil profiles.

In order to confirm the applicability of magnetic susceptibility measurements to identify the layers of technogenic origin, basic statistical analysis has been done, in which the relationship between the distribution of magnetic susceptibility and all analysed chemical parameters was determined. Values of Pearson's correlation coefficient were significant for the measurement of electrical conductivity ($r = 0.7433$) (Fig. 3) and the fraction of 0.02–0.005 ($r = 0.4853$), 0.005–0.002 ($r = 0.5590$) and <0.002 mm ($r = 0.4738$) (Fig. 4).

Measurements of the magnetic susceptibility, characterized by high sensitivity and precision, may be used to identify changes caused by the presence of even a small amount of ferromagnets of technogenic origin (Strzyszcz 1993; Hanesch and Schloger 2005; Zan et al. 2015; Wawer et al. 2015). This feature enables the identification of genetic horizons, where, due to absorption of ferromagnetics, magnetic properties undergo changes. Thus, there is the possibility of a precise delimitation of individual horizons in the soil profile.

5 Conclusions

1. Soils formed after the construction of the Vistula River Cross-Cut are characterized by a very dynamic development. This is indicated primarily by their morphological structure—accumulation of natural mineral deposits (alluvial, aeolian and marine) mixed with technogenic ones (excavated by man during hydrotechnical works) in the layers of relatively small thickness with clear boundaries between particular genetic horizons.
2. Measurements of magnetic susceptibility are a precise tool for supporting classification of genetic horizons existing in the soil profile, with particular emphasis on the occurrence of discontinuities of technogenic origin. Due to a very high sensitivity of magnetic measurements it is possible to determine sites with the accumulation of even very small amount of ferromagnets.
3. The results of the study showed a high correlation between the values of electrical conductivity, the content of the fraction with a diameter of <0.02 mm and the measurement of magnetic susceptibility.

Acknowledgements This study was financed by the Polish National Science Centre (Grant DEC 2012/07/B/ST10/04080, 2013–2016).

References

Dearing J (1999) Environmental magnetic susceptibility: using the Bartington MS2 System. Chi Publishing, Keniloworth, UK

Dearing J, Dann R, Hay K, Lees J, Loveland P, Maher B, O'Grady K (1996) Frequency-dependent susceptibility measurements of environmental materials. Geophys. J. Int. 124:228–240

Duan XM, Hu SY, Yan HT, Blaha U, Roesler W, Appel E, Sun WH (2010) Relationship between magnetic parameters and heavy element contents of arable soil around a steel company, Nanjing. Sci. China (Earth Sci) 53:411–418

Gelisli K, Aydin A (1998) Investigation of environmental pollution using magnetic susceptibility measurements. European J. Environ. Eng. Geophysics 3:53–61

Guidelines for Soil Description (2006) Food and Agriculture Organization of the United Nations, Rome

Han GZ, Zhang GL (2013) Changes in magnetic properties and their pedogenetic implications for paddy soil chronosequences from different parent materials in south China. Eur. J. Soil Sci. 64:435–444

Hanesch M, Scholger R (2002) Mapping of heavy metal loadings in soils by means of magnetic susceptibility measurements. Environ. Geol. 42:857–870

Hanesch M, Scholger R (2005) The influence of soil type on the magnetic susceptibility measured throughout soil profiles. Geophys. J. Int. 161:50–56

Hanesch M, Rantitsch G, Hemetsberger S, Scholger R (2007) Lithological and pedological influences on the magnetic susceptibility of soil: their consideration in magnetic pollution mapping. Sci. Total Environ. 382:351–363

Hulisz P (2013) Genesis, properties and systematics position of the brackish marsh soils in the Baltic coastal zone (Geneza, właściwości i pozycja systematyczna marszy brakicznych w strefie oddziaływania wód Bałtyku). Rozprawy habilitacyjne. Wyd. UMK: 137 pp. (in Polish)

Hulisz P, Michalski A, Dąbrowski M, Kusza G, Łęczyński L (2015) Human-induced changes in the soil cover at the mouth of the Vistula River Cross-Cut. Soil Sci. Ann. 66(2):67–74

Kapička A, Petrovský E, Jordanova N, Podrázský V (2001) Magnetic parameters of forest top soils in Krkonose Mountais, Czech Republik. Phys. Chem. Earth 26(11–12):917–922

Magiera T, Strzyszcz Z, Kapička A, Petrovský E (2006) Discrimination of lithogenic and anthropogenic influences on topsoil magnetic susceptibility in Central Europe. Geoderma 130:299–311

Magiera T, Strzyszcz Z, Rachwal M (2007) Mapping particulate pollution loads rusing soil magnetometry in urban forests in the Upper Silesia Industrial Region, Poland. For. Ecol. Manag. 248:36–42

Makowski J (1995) The hundredth anniversary of the creating the Vistula Cross-Cut 1895–1995 (Setna rocznica wykonania Przekopu Wisły 1895–1995). IBW PAN, Gdańsk in Polish

Oldfield F (1991) Environmental Magnetism—a personal prospective. Quat. Sci. Rev. 10:73–85

Robakiewicz M (2010) Vistula River mouth—history and recent problems. Arch. Hydro-Eng. Environ. Mech. 57(2):155–166

Strzyszcz Z (1993) Magnetic susceptibility of soils in the areas influenced by industrial emissions. In: Schulin R, Sesaules A, Webster R, Steiger BV (eds) Soil monitoring. Birkhäuser Verlag, Basel, pp 255–269

Strzyszcz Z, Magiera T (1998) Magnetic susceptibility and heavy metals contamination in soils of southern Poland. Phys. Chem. Earth 23:1127–1131

Strzyszcz Z, Rachwał M (2008) Changes in magnetic susceptibility of forest soils along the west and south border of Poland. Arch. Environ. Prot. 34(1):71–79

Strzyszcz Z, Magiera T, Bzowski Z (1994) Magnetic susceptibility as indicator of soils contamination in some regions of Poland. Soil Sci. Ann. 44:85–93

Thompson R, Oldfield F (1986) Environmental magnetism. Allen and Unwin, London, UK

Wawer M, Magiera T, Ojha G, Appel E, Bućko MS, Kusza G (2015) Characteristics of current roadside pollution using test-monitoring plots. Sci. Total Environ. 505:795–804

Wróblewski R, Rudowski S, Gajewski Ł, Sitkiewicz P, Szefler K, Kałas M, Koszałka J (2015) Changes of the Vistula River external delta in the period of 2009–2014. Bull. Marit. Inst. Gdańsk 2(1):16–22
Zan J, Fang X, Yan M, Zhang Z, Zhang D (2015) Regional variations in magnetic properties of surface sediments in the Qaidam Basin and their paleoenvironmental implications. J. Appl. Geophys. 122:86–93
Zawadzki J, Magiera T, Fabijańczyk P (2009) Geostatistical evaluation of magnetic indicators of forest soil contamination with heavy metals. Studia Geophys et Geod 53:133–149
Zawadzki J, Fabijańczyk P, Magiera T, Strzyszcz Z (2010) Study of litter influence on magnetic susceptibility measurements of urban forest topsoils using the MS2D sensor. Environ. Earth Sci. 61:223–230

Magnetic Susceptibility of Sediments as an Indicator of the Dynamics of Geomorphological Processes

Elżbieta Król and Piotr Szwarczewski

Abstract This paper presents the results of measurements of the magnetic susceptibility of samples taken in the profiles of Holocene sediments. Locations from which samples were analyzed were previously investigated with various methods (such as geological, geomorphological, sedimentological, geochemical and pollen). Most of the examined sites have a well-known geological structure, the origin and age of existing forms. The sediments recognized with the drillings represent differentiated sedimentary environments (e.g., slopes, fluvial), differentiated facies (mineral and mineral-organic deposits, peat, carbonaceous tufa) and come from different types of the landscape: from Polish loess plateaus and associated valleys bottoms of Ponidzie to Middle Poland plains with dunes and alluvial fans occurring on the terrace in the Vistula River valley. Analysis of changes in magnetic susceptibility in selected well-recognized (i.e. of known age, origin and sedimentological features) vertical profiles makes it possible to determine the relationship between the measured value of magnetic susceptibility and environmental changes that have taken place in the past. Knowledge of these relationships allows to read the events in the development of the natural environment of the studied areas and in the case of possession of radiocarbon dating in the reference profiles it allows to make a vertical correlation with the neighboring sediment profiles of unknown age.

Keywords Magnetic susceptibility · Human impact · Radiocarbon dating · Environmental changes

E. Król (Deceased)
Department of Earth Magnetism, Institute of Geophysics Polish Academy of Sciences, ks. Janusza 64, 01-452 Warsaw, Poland
e-mail: elakrol@igf.edu.pl

P. Szwarczewski (✉)
Department of Geomorphology, Institute of Physical Geography, Faculty of Geography and Regional Studies, University of Warsaw, Krakowskie Przedmieście 30, 00-927 Warsaw, Poland
e-mail: pfszwarc@uw.edu.pl

M. Jeleńska et al. (eds.), *Magnetometry in Environmental Sciences*,
GeoPlanet: Earth and Planetary Sciences, DOI 10.1007/978-3-319-60213-4_6

1 Introduction

Magnetic susceptibility is one of magnetic parameters widely used in studies of sedimentary rocks for analysis of influence of climate and anthropopression on such types of young (Quaternary) sediments as loess, lacustrine sediments and for study of pedogenenic processes in soils (e.g., Badura et al. 2003; Bidegain et al. 2009; Petrovsky et al. 2001; Williamson et al. 1998; Zolitschka et al. 2003).

When the accumulated sediments have similar characteristics (sedimentological, genetic), variation of magnetic susceptibility may be treated in some cases as an additional feature that reflects the changes taking place in the natural environment under the influence of climate and/or human economic activity.

Magnetic susceptibility of sediment is used for a long time in studies of environmental changes but mostly it is used in the analysis of the loess and lacustrine sediments (e.g., Badura et al. 2003; Bidegain et al. 2009; Petrovsky et al. 2001; Williamson et al. 1998; Zolitschka et al. 2003).

2 Study Area and Methods

This paper presents the results of measurements of the specific magnetic susceptibility of samples taken in the profiles of the Holocene sediments. Measurements have been conducted using Kappa-bridge KLY-2 (made by "Geofizyka"—Brno, Czech Republic). The maximal sensibility of the KLY-2 bridge is 10^{-8} for measurements of bulk magnetic susceptibility κ. The measured values of κ of dried, loose specimens have been counted over for the mass specific magnetic susceptibility χ thanks to known mass of every specimen and taking into account that the bulk magnetic susceptibility κ in the KLY-2 bridge is calibrated for specimens of known, constant volume (10 cm^3).

Locations from which samples come were analyzed previously by various methods (such as geological, geomorphological, sedimentological, geochemical or pollen) and the results were partially published (e.g. Szwarczewski 2007, 2009; Szwarczewski et al. 2008).

The examined sites have in majority a well-known geological structure and the origin and age of existing forms. The sediments recognized by drillings represent a variety of sedimentary environments (e.g., slope, fluvial) and they are facially differentiated (i.e., mineral and mineral-organic deposits, peat, calcareous tufa). They come from different types of the landscape (Fig. 1), i.e., Polish loess plateaus and associated dry and river valleys of Ponidzie (Chroberz and Pełczyska sites), the Mid-Poland plains with dunes and alluvial fans occurring on one of the terraces of the Vistula River valley (Prague terrace) and the peat-dome with calcareous tufa in the vicinity of Kuźnica (Sokólskie Hills region, NE Poland).

So this is an attempt to verify whether there is a response of the environmental changes or human impact in the magnetic susceptibility values along the vertical

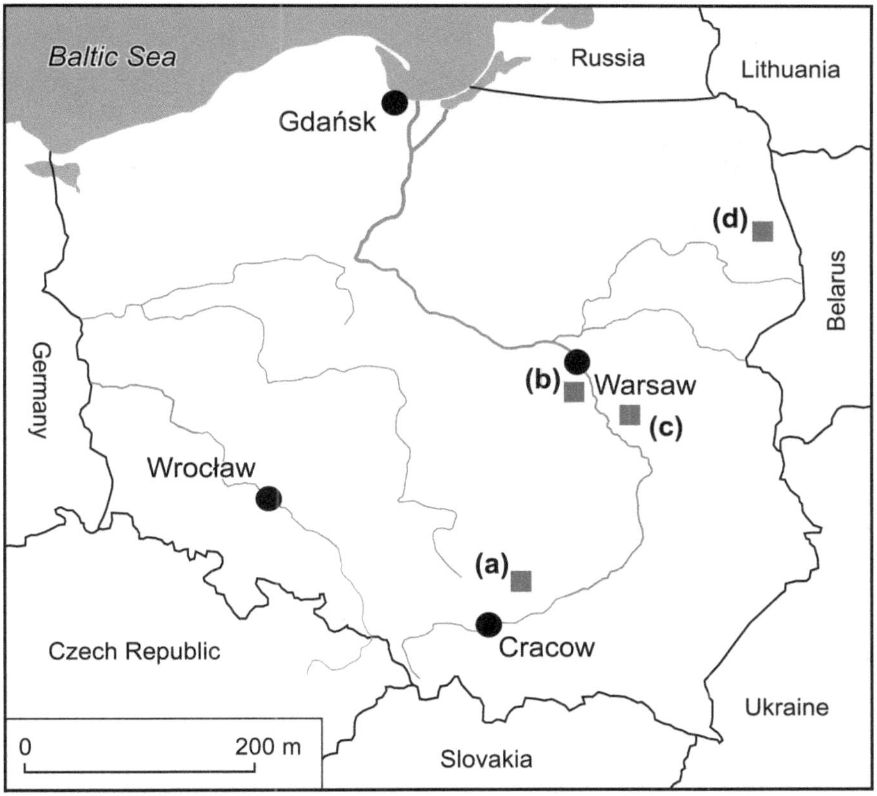

Fig. 1 Location of the study areas. **a** Mozgawa-Kozubów and Pełczyska sites, **b** Ursynów Escarpment sites, **c** Glina site, **d** cupola spring mire in Kuźnica

profiles. The samples tested in each of the profiles were generally collected every 5–20 cm.

Most of the profiles were absolutely dated using radiocarbon method and their results (uncalibrated age) are shown in the drawings (Figs. 2, 3, 4, 5 and 6).

The observed magnetic susceptibility depends on the content of the sum of ferromagnetic (sensu lato), paramagnetic and diamagnetic minerals present in every specimen (Evans and Heller 2004). It is obvious that the content of such minerals in the samples can be very changeable and the observed magnetic susceptibility in sedimentary rocks depends not only on their primary accumulation during sedimentation but also on the further syn- and post-sedimentary processes in such rocks.

They are mainly connected with chemical reactions of oxidation and reduction. The rate of such reactions depends usually on temperature and humidity (the straight influence of existing climatic conditions) and migration of solutions in sediments—which is connected with selective enrichment of some levels of studied rocks by chosen magnetic fractions.

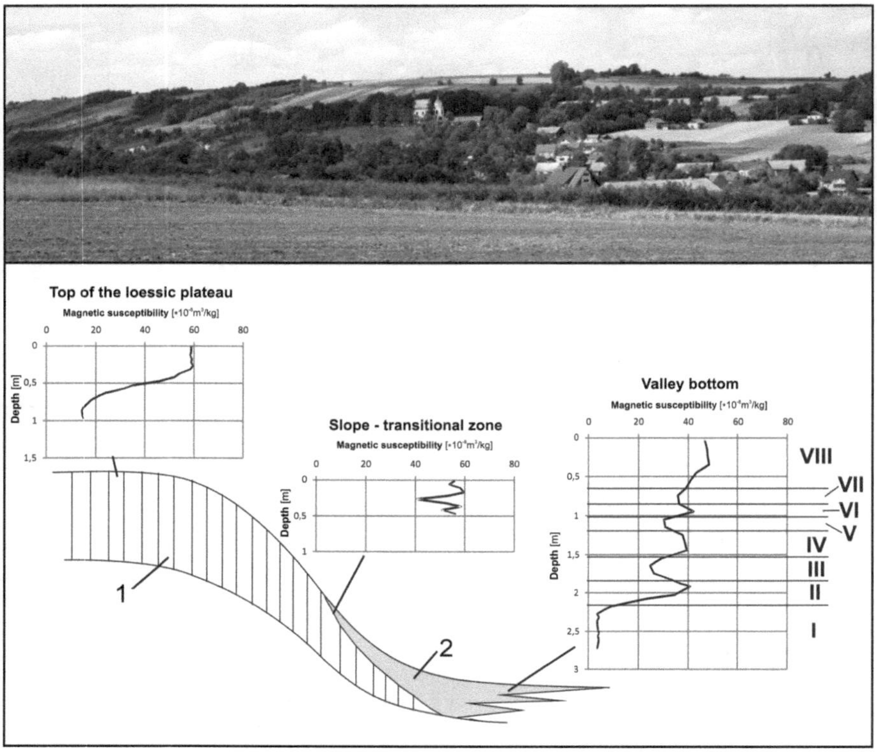

Fig. 2 Schematic geological structure of the slope on loess plateau near Pełczyska and measured values of magnetic susceptibility in selected profiles. *1* loess (Pleistocene), *2* loessic delluvia (Holocene). *I–VIII* phases of accumulation of sediments related to climate change and development of the settlements in the study area. The attached photo presents a view for the study site

The additional anthropogenic influence on the rate and direction of such reactions should be also taken into account in the case of increasing content of organic matter, the use of mineral fertilizers in the surface layers of soils and increasing supply from atmospheric pollution. The local fires can also change the mineral content of ferromagnetic fraction in soils.

All these factors can change the structure of existing ferromagnetic minerals and cause the origin of new portions of ferromagnetic minerals on the cost of Fe present in paramagnetic minerals—the main components of sedimentary rocks.

The physical processes such as diagenesis, erosion, floods, neo-tectonic movements and rebuilding of slopes also have an influence on the properties of studied sediments, which is reflected in vertical changes of magnetic susceptibility.

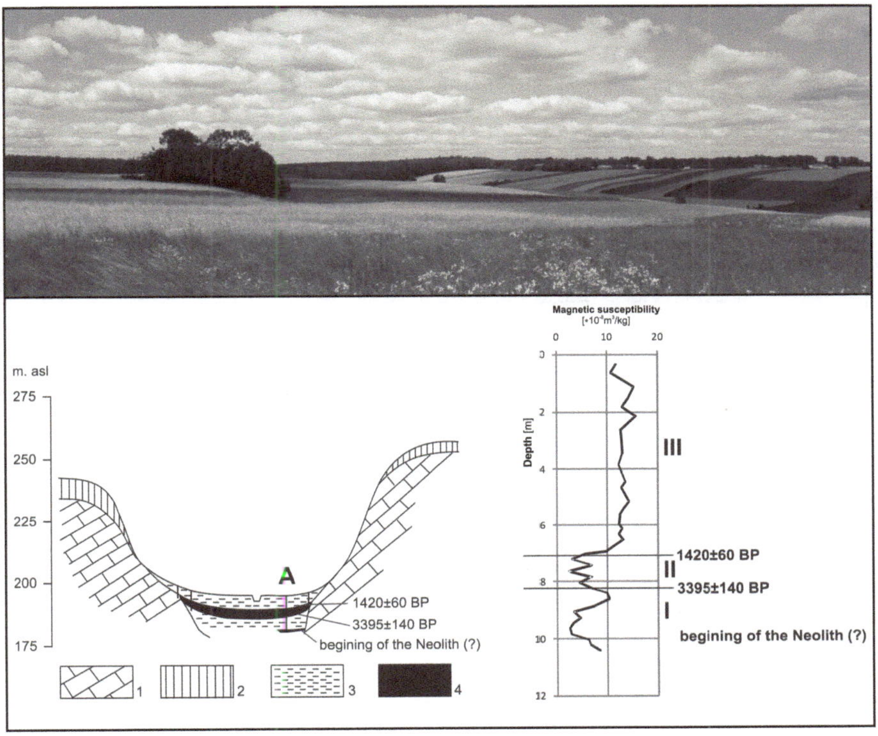

Fig. 3 Geology of the Mozgawka Valley in the vicinity of Mozgawa-Kozubów and measured values of MS in profile A. *1* marls (Cretaceous), *2* loess (Pleistocene), *3* loessic delluvia (Holocene), *4* peat. *I–III* phases of accumulation of sediments related to climate change and development of the settlements in the study area. The attached photo presents a view for the study site

3 Results

Facial diversification of the sediments occurring on attractive, used for some 5000 years for agricultural purposes, slopes of loess plateau in the vicinity of Mozgawa-Kozubów and Pełczyska and in the neighbouring bottoms of valleys or local depressions is a result of the primeval/historic slope processes due to the variable in time economic activity of the man (Szwarczewski 2009). It is also worth stressing that the bottoms of many valleys and local depressions during the appearance of the first farmers and breeders in the loessic regions of southern Poland were in many cases about 10 meters lower than at present (eg. Fig. 3).

Because loess sediments are very homogeneous and relatively well-sorted the differentiation of magnetic susceptibility values occurring in the areas of loess plateaus and highlands should be interpreted as a record of the variable rate of

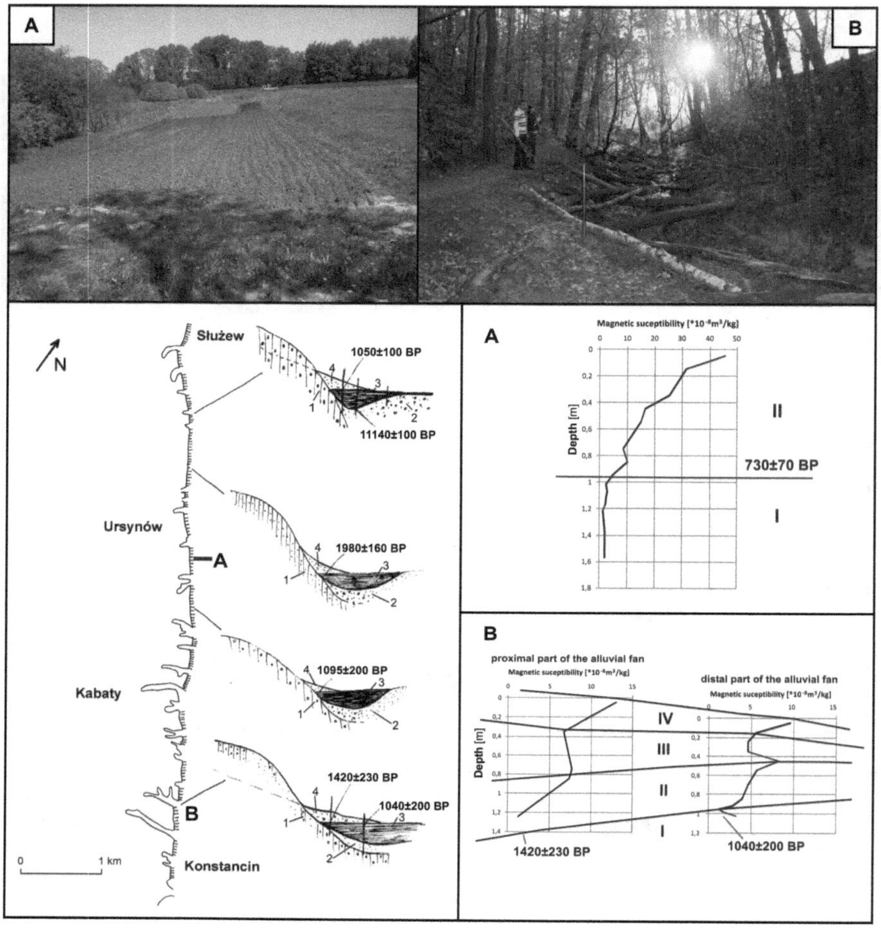

Fig. 4 Geomorphological sketch of Ursynów scarp and schematic geological cross-sections. *A*: *I–II* phases of accumulation of sediments related to development of the settlements in the study area. *B*: *I–IV* phases of accumulation of sediments related to development of alluvial fan. The attached photos presents a view for the study site (*A* and *B*)

delivery of material from erosion of slopes and parts of plateaus. The results of radiocarbon datings together with magnetic susceptibility data indicate that the acceleration of the erosional processes was related to settlement development and introduction of agriculture and breeding activity. It was also important the development of metallurgy and industrial revolusion. The profiles located in Ponidzie near Mozgawa-Kozubów (Mozgawka valley) (Fig. 2) and Pełczyska (Fig. 3) have documented the described changes quite satisfactorily.

The development of human settlement in the vicinity of Ursynów Escarpment (in Warsaw) has been connected with the appearance of many erosional forms

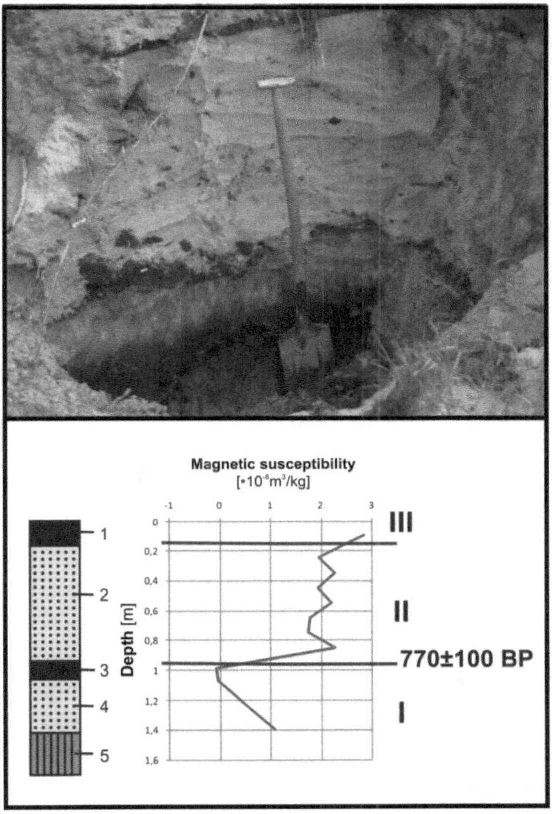

Fig. 5 Schematic geological structure of the profile of eolian deposits in Glina near Otwock and measured values of magnetic susceptibilty. The attached photo present the structure of the analysed outcrop. *1* present top soil, *2* massive eolian deposits, *3* burried top soil horizon (radiocarbon dated), *4* fluvial/eolian sands, *5* glacial till *I–III* phases of accumulation of sediments related to the land use changes and development of the settlements in the study area

(mostly gullies), which had cut morainic plateau adjoining to the Vistula river valley (Fig. 4).

Results of radiocarbon dating of sediments which have been accumulated beneath the slopes of Ursynów Escarpment and at the mouths of the dry valleys and gullies correlate very well with the time of appearance of human settlement in this area (especially the late Roman and Early Medieval periods). The increasing economical activity of population in this part of Warsaw is documented by the increasing values of magnetic susceptibility in the upper parts of studied profiles—higher values of the MS in the profile are corelated with the higher Fe, Mn and trace elements contents that were due to the industrial development, especially since the 2nd part of 19 century (Szwarczewski 2007). The value of this parameter in the sediments taken from distal and proximal parts of alluvial fans let us correlate the sediments of the same age and the changes of magnetic susceptibility along the following profiles brings the information about their development in time (Fig. 4).

The very small changes of magnetic susceptibility measured in the massive, aeolian sands and their different character in comparison with the sediments present above and beneath this layer indicate that the process of aeolian accumulation was very fast, almost with no stops. In addition, the radiocarbon age of charcoal, which has been found in this site together with the age of hummic horizon of fossil soil, confirms that this morphological process was connected with agriculture development after deforestation in the early Middle Ages (Fig. 5). This deposition of aeolian, unfertile sands prevented further agricultural use of this land.

The measurements of magnetic susceptibility of sediments which are present in the peat-dome with calcareous tufa in the vicinity of Kuźnica (Sokólskie Hills region, NE Poland) bring the interesting palaeogeographic information about this area. The wet climate and the increase of denudation caused the appearance of calcium carbonate sediments (of tufa type) during the great part of Early Holocene and Atlantic periods. In addition, bog iron ores (with very high magnetic susceptibility) have appeared in a part of this site. Their existence is determined by the more wet period during the Atlantic part of the Holocene (Fig. 6).

4 Discussion

Accumulation of sediments of similar geological origin and character causes the similar character of their mineral structure and, as a result, a similar range of their magnetic susceptibility. However, the observed changes of magnetic susceptibility can sometimes be treated as results of additional natural and artificial factors such as an influence of climate changes or the result of anthropopression. It is extremely visible in the subordinate landscapes (valley bottoms, base of the slopes, lake sediments etc.) where vertical variation of the magnetic suseptibility reflects the land use changes and actual human activity—from gathering and hunting, via agriculture and shepherding to the industrial one. The changes of MS values in the sediments from slope erosion deposited in the depressions can be treated as the straight forward geophysical evidences for the phases of the human economic activity that took place in the past. Some of mentioned above phases were better or worse marked in the presented sites and profiles (Figs. 2, 3, 4, 5 and 6).

The separation of these two main factors which influenced the environmental changes during the Holocene on the basis of analysis of one proxy, such as magnetic susceptibility, is very difficult and sometimes impossible.

The additional magnetic parameters: Curie temperatures of magnetic minerals present in the following layers of studied sediments and parameters of magnetic hysteresis loop can elucidate the process of their formation. The susceptibility measured for specimens magnetized by stable and changeable magnetic field (the so-called anhysteretic magnetic remanence—ARM), measurements of saturation magnetization and frequency dependent susceptibility let us determine percentage content of natural magnetic minerals in comparison with the new magnetic minerals, which appear in sediments thanks to the above-mentioned chemical reactions.

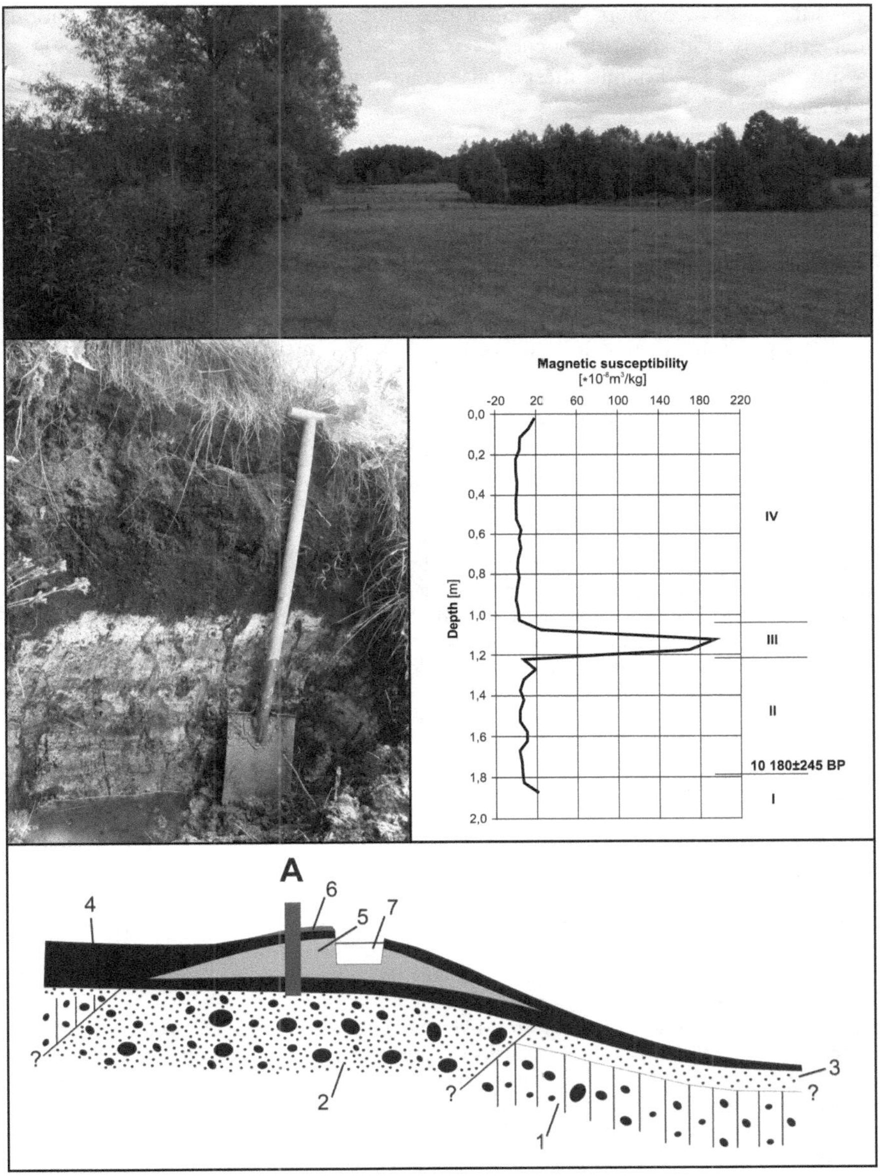

Fig. 6 Schematic geological cross-section of cupola spring mire in Kuźnica (after Szwarczewski et al. 2008) and measured values of magnetic susceptibility in profile A. *1* boulder clay (Pleistocene), *2* sand and gravel with boulders (Pleistocene), *3* sand and gravel (Holocene), *4* peat (Holocene), *5* calcareous tufa, *6* overburden anthropogenic material (carbonate sediments and peat), *7* pond. The attached photographs present a view of the site (top one) and the structure of the analysed outcrop (bottom one)

The differences of domain structure of magnetic minerals and in the dimension of their grains decide about their coercion.

In the studied sediments, these analyses have not been done yet, but are planned in future.

5 Conclusions

An analysis of changes in magnetic susceptibility in selected well-recognized (age, origin and sedimentological features) vertical profiles makes it possible to determine the relationship between the measured value of magnetic susceptibility and environmental changes that have taken place in the past. Knowledge of these relationships allows reading the events in the development of the natural environment of the studied areas and in the case of possession of radiocarbon dating in the profiles of the reference it allows to make a correlation with the neighboring sediment vertical profiles of unknown age.

Acknowledgements The research was financed by the grants: (1) from the Ministry of Science and Higher Education—Selected records of human impact in the evolution of the natural environment of Polish lowlands—primeval, historical and contemporary aspects (nr N306 013 31/0757, 2006-2010) and (2) from National Science Centre—Archaeological, archaeobotanical and palaeoenvironmental investigations in the western part of the Nida Basin (nr 2013/11/B/HS3/03822, 2014-2017).

References

Badura J, Bluszcz A, Brizova E, Ciszek D, Issmer K, Moska P, Pawlyta J, Pazdur A (2003) Holoceńskie zmiany środowiska przyrodniczego na przykładzie rozwoju stożka aluwialnego w rejonie Wzgórz Korzuchowskich (SW Polska) (Holocene environmental changes on the example of alluvial fan development in the region of Korzuchowskie Hill (SW Poland). In: Waga JM, Kocel K (eds) Człowiek w środowisku przyrodniczym—zapis działalności. UŚ Sosnowiec, 5–15. (in Polish)

Bidegain JC, Rico Y, Bartel A, Chaparro MAE, Jurado S (2009) Magnetic parameters reflecting pedogenesis in Pleistocene loess deposits of Argentina. Quatern. Int. 209:175–186

Evans ME, Heller F (2004) Environmental magnetism—principles and applications of enviromagnetics. Academic Press, San Diego, p 299

Petrovsky E, Kapicka A, Jordanova N, Boruvka L (2001) Magnetic properties of alluvial soils contaminated with lead, zinc and cadmium. J. Appl. Geophys. 48:127–136

Szwarczewski P (2007) Etapy rozcinania Skarpy Ursynowskiej (Warszawa) (Phases of the Ursynow Scarp (Warsaw, Poland) development). Prace Instytutu Geografii AŚ w Kielcach 16:157–171

Szwarczewski P, Dąbski M, Pawłowski K (2008) Kopuła torfowo-martwicowa w Kuźnicy - wstępne wyniki badań (Peat and calcareous tufa dome in Kuznica area—preliminary results of investigation). Geneza, litologia i stratygrafia utworów czwartorzędowcyh. UAM, Poznań, pp 123–124

Szwarczewski P (2009) The formation of deluvial and alluvial cones as a consequence of human settlement on a loess plateau: an example from the Chroberz area (Poland). Radiocarbon 51(2):445–455

Williamson D, Jelinowska A, Kissel C, Tucholka P, Gibert E, Gasse F, Massault M, Taieb M, Van Campo E, Wieckowski K (1998) Mineral-magnetic proxies of erosion/oxidation cycles in tropical maar-lake sediments (Lake Tritrivakely, Madagascar): paleoenvironmental implications. Earth Planet. Sci. Lett. 155:205–219

Zolitschka B, Behre K-E, Schneider J (2003) Human and climatic impact on the environment as derived from colluvial, fluvial and lacustrine archives—examples from the Bronze Age to Migration Period, Germany. Quatern. Sci. Rev. 22:81–100

The Impact of Grain Size Composition and Organic Matter Content on Magnetic Susceptibility of Anthropogenically Transformed Bottom Sediments, as Exemplified by the Former Naval Harbour in Hel

Leszek Łęczyński, Żaneta Kłostowska, Grzegorz Kusza, Tadeusz Ossowski, Bartłomiej Arciszewski and Radomir Koza

Abstract The article presents the results of an analysis of the surface layer of bottom sediments in the naval harbour in Hel. During field work conducted in June 2013, 54 sediment samples were collected. The aim of the study was to examine magnetic susceptibility of harbour sediment as a parameter determining the degree of anthropopressure. As part of laboratory tests, grain size analysis was performed in order to determine organic matter content and magnetic susceptibility of the sediment. The study also measured the impact of environmental factors on fluctuations in the other studied parameters. The results demonstrate increased dynamics of sedimentary environment in the open part of the naval harbour, better sediment sorting and reduced percentage share of <0.063 mm grain size fraction. Variability in magnetic susceptibility values measured in isolated parts of the harbour correlates with increased percentage share of the fine fraction and LOI%. A significant impact on the accumulation of pollutants in surface sediments is exerted by the harbour's closed hydrodynamic conditions.

L. Łęczyński · Ż. Kłostowska (✉)
Laboratory of Applied Geology, Institute of Oceanography, University of Gdańsk, Piłsudskiego 46, 81-378 Gdynia, Poland
e-mail: zaneta.klostowska@phdstud.ug.edu.pl

G. Kusza
Department of Land Protection, University of Opole, Oleska 22, 45-052 Opole, Poland

T. Ossowski
Department of Analytical Chemistry, Faculty of Chemistry, University of Gdańsk, Wita Stwosza 63, 80-308 Gdańsk, Poland

B. Arciszewski · R. Koza
Institute of Oceanography, Hel Marine Station, University of Gdańsk, Morska 2, 84-150 Hel, Poland

M. Jeleńska et al. (eds.), *Magnetometry in Environmental Sciences*,
GeoPlanet: Earth and Planetary Sciences, DOI 10.1007/978-3-319-60213-4_7

Keywords Naval harbour · Bottom sediments · Magnetic susceptibility · Anthropopressure

1 Introduction

Bottom sediments in harbours constitute an interesting object of study due to the transformations they undergo under the influence of anthropogenic factors, which cause irreversible changes in the sedimentary environment. Both the military function of the facility and the processes occurring in harbour areas affect physical and chemical parameters of sediments.

The former naval harbour in Hel is located in the southern part of the Hel Peninsula, close to the eastern coast of the Bay of Puck. Construction of the harbour started in 1931 in the vicinity of the so-called Old Hel, which was the first local settlement. The harbour was built by a Polish-French consortium and the design was created by Włodzimierz Szawernowski. Till the year 1929 a wet dock 400 × 300 m in size was constructed along with a 45 m-wide quay. Then, to meet the needs of the nearby naval base, harbour infrastructure was developed. About 1.5 km north of the harbour, an underground power station was built and ammunition, torpedo and mine bunkers were constructed in a neighbouring forest (Pasecki 2010). After WWII the harbour was modernised. A new wet dock with an entrance head was built, as a result of which the harbour more than doubled in size towards the fishing port.

The harbour was used by the Polish Navy till 2008. Currently, it is being managed by the Military Property Agency and used as a base for naval ships taking part in military manoeuvres. There are plans to make the area available for seamen as well as for hotel and residential investments. This is why it is important to determine the exact degree to which harbour sediments have been subjected to anthropopressure.

Initial evaluation was based on the measurements of selected magnetic parameters of sediments, which were correlated with the content of organic matter and grain size (<0.063 grain size fraction) of the surface layer of bottom sediment.

2 Materials and Methods

2.1 Field Work

Research area was divided into two parts—inner and outer—according to guidelines on hydrotechnical constructions. As part of the research, 50 points were selected within the harbour and 4 reference points along the coastal infrastructure of the Hel Peninsula (Fig. 1). Different parts of the harbour vary in depth. The inner

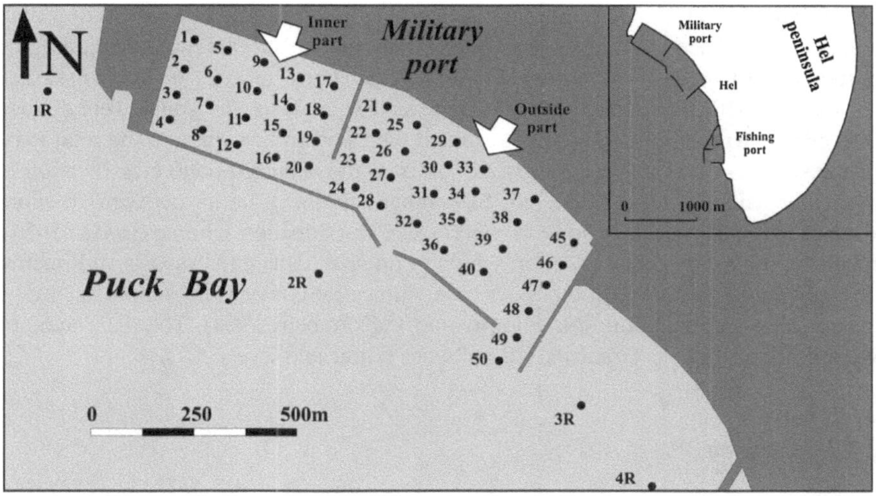

Fig. 1 Location of sediment sampling points in the harbour and in the Bay of Puck

part is between 6 and 11 m deep, while the outer part is more shallow, with the depth ranging from 4 to 8 m. Sediment samples were collected in July 2013 from a hydrographic vessel belonging to the Marine Station of the Institute of Oceanography, University of Gdańsk. The surface layer of the sediment was sampled with the use of Van Veen Grab Sampler, used for unconsolidated sediments with disturbed structure, such as sand, silt or clay (Piekarek-Jankowska and Wojcieszek 2010). Collected samples were put in polythene bags and conserved by freezing until laboratory analyses were carried out.

2.2 *Laboratory Analyses*

The analyses of bottom sediment samples were conducted in the Department of Marine Geology, University of Gdańsk, and the autonomous Department of Land Protection, University of Opole.

2.2.1 Analysis of Grain Size Composition

In order to determine the granulometric composition of the sediment material, the samples were subjected to sieve analysis. The sediments were dried in 60 °C for 24 h, until their mass ceased to change (Winters et al. 2000). After an initial macroscopic analysis, appropriate test portions were weighed out for analysis. The samples were put in Fritsch Analysette 3 vibratory sieve shaker, which consists of a set of normalised test sieves (aperture size: 2.0, 1.0, 0.5, 0.25, 0.125 and 0.063 mm) and a shaker. Once the samples had been sifted, each of the fractions remaining on particular test sieves was weighed with the accuracy of 0.01 g (Łęczyński and Szymczak 2010). Based on the results of grain size analysis, the grains were divided into groups within particular size range and their percentage share in the total mass of the sediment was calculated. On the basis of the acquired data, classification of particular samples was conducted. Size ranges defining limits between fractions were established according to Udden grain size classification scheme (Udden 1914), modified by Wentworth (1922), in which grain size (d) is expressed in millimetres or logarithmic units (φ). In the latter case, numerical conversion is done according to the following equation: $\varphi = -\log_2 d$ (mm) (Krumbein 1964). The data obtained were processed using Gradistat 8.0 software (Blott and Pye 2001).

2.2.2 Humidity

Sediment humidity was measured on the basis of the quantitative relation [%] between the amount of water contained in a particular sample before and after drying it (Myślińska 2010). Porcelain crucibles, dried and cooled to room temperature, were weighed with the accuracy of 0.0001 g. Next, 20–30 g of sediment were put in each of them, after which the crucibles were weighed with the same accuracy as before (Winters et al. 2000). For each of the sediment samples, the analysis was carried out three times.

2.2.3 Organic Matter Content (LOI)

Organic matter content [%] in the sediment was measured by the loss on ignition method (LOI) in 550 °C, a temperature accepted as optimal for Baltic sediments (Ciborowski 2010). Test portions (10 g) of previously dried and homogenised samples were put in porcelain crucibles weighed with the accuracy of 0.0001 g. The samples were ignited until their mass ceased to change. Then the loss of mass was calculated and expressed as percentage of initial value. For each of the samples, the analysis was carried out three times.

2.2.4 Magnetic Susceptibility

In order to measure magnetic susceptibility, sediment material was air dried. Samples were put into containers and weighed with the accuracy of 0.001 g. Magnetic susceptibility measurements were conducted with the use of Barington MS2B sensor at two different frequencies: low 0.456 kHz (klf) and high 4.65 kHz (khf). The measurements were repeated three times. On their basis, volume magnetic susceptibility (κ) was calculated, which—after taking into account the mass and density of each sample (ρ)—made it possible to determine mass (specific) magnetic susceptibility: $\chi = \kappa/\rho$ ($\times 10^{-8}$ m^3 kg^{-1}).

Methods of measuring magnetic susceptibility with the use of Bartington MS2B sensor are presented in various papers (Magiera et al. 2002; Strzyszcz and Rachwał 2010; Zawadzki et al. 2010, 2012; Gołuchowska et al. 2012).

3 Results and Discussion

In lithological sense, harbour sediments are made up mainly of medium-grained sand. The analysed bottom sediments were characterised by the highest content of 0.25 mm fraction, which constituted between 14 and 75% of all samples (Fig. 2). The sediments contained in the 50 harbour samples and 4 reference samples were classified as medium-grained sand. In the analysed area, most grains fell within a narrow range of diameters (from 0.5 to 0.125 mm), which may mean that the sediments formed in a stable environment. The bottom of the harbour also contained some gravel (grains larger than 2 mm), but its amount was very small. The highest content of gravel fraction (grains 2–4 mm in diameter) was found in sample no. 4 and amounted to 7.5%.

The analysis of <0.063 mm grain size distribution in harbour sediments shows that it varies with location. In the inner part, which lacks direct water exchange with the bay, <0.063 mm fraction amounts to approximately 7% in the granulometric distribution of the analysed sediment. In the sediment collected in the outer part, on the other hand, the percentage share of <0.063 mm fraction was calculated to be

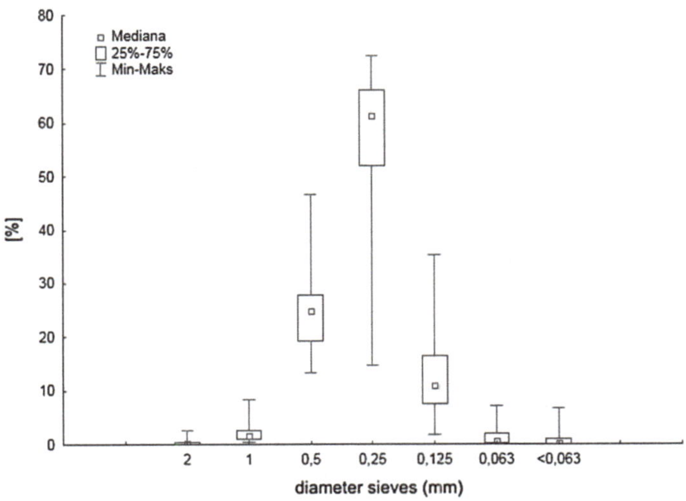

Fig. 2 Percentage share of particular grain-size fractions in harbour sediments

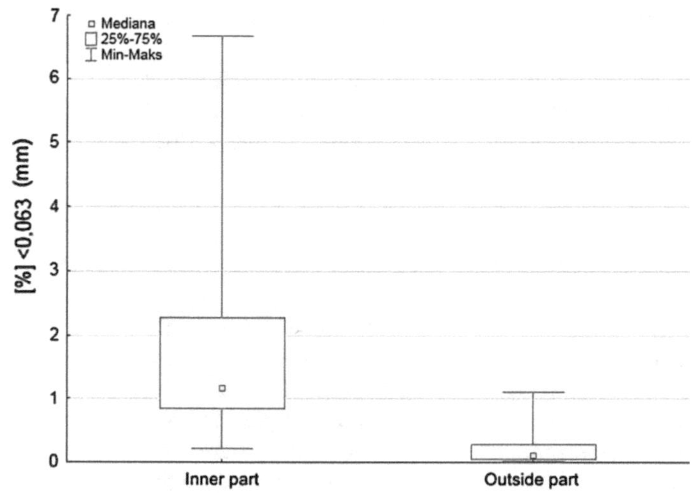

Fig. 3 Minimum and maximum values as well as the mean percentage share of <0.063 mm fraction in the sediments collected within the inner and outer parts of the harbour

below 1%, which is caused by increased environmental dynamics at the entrance to and exit from the harbour (Fig. 3).

Organic matter content in the analysed sediment samples varies according to the location of sampling points and—at the same time—their distance from hydrotechnical buildings. Organic matter content in surface layer bottom sediments did not exceed 0.5% in 30 sampling points (Fig. 4). In the inner part of the harbour, organic matter content reached 4% (Fig. 5) and thus was higher than in the outer part, where it did not exceed 1.1% (Fig. 6).

Likewise, the measured magnetic susceptibility values varied with location within the harbour.

In this study, magnetic susceptibility measurements were treated as an indicator of changes triggered by anthropogenic transformations of harbour sediments. The analysis also took into account the distribution of granulometric diversity values and organic matter content in the analysed samples as opposed to the reference samples collected outside the harbour. Magnetic susceptibility values (χ) for the sediments from the inner part ranged between 0 and 10×10^{-8} m^{-3} kg^{-1}, with the mean of 4.68×10^{-8} m^{-3} kg^{-1} (Fig. 7). The mean value of magnetic susceptibility for the outer part was 2.52×10^{-8} m^{-3} kg^{-1}. Among the analysed sediment samples, the majority had susceptibility values (χ) ranging between 0 and 5×10^{-8} m^{-3} kg^{-1} (Fig. 8). Magnetic susceptibility values for reference samples fluctuated within the range of 0×10^{-8} m^{-3} kg^{-1} and 1.1×10^{-8} m^{-3} kg^{-1}, and were 0 for points R1 and R3.

Statistical tests revealed significant correlations between the value of mass magnetic susceptibility and the percentage share of <0.063 mm grain size fraction

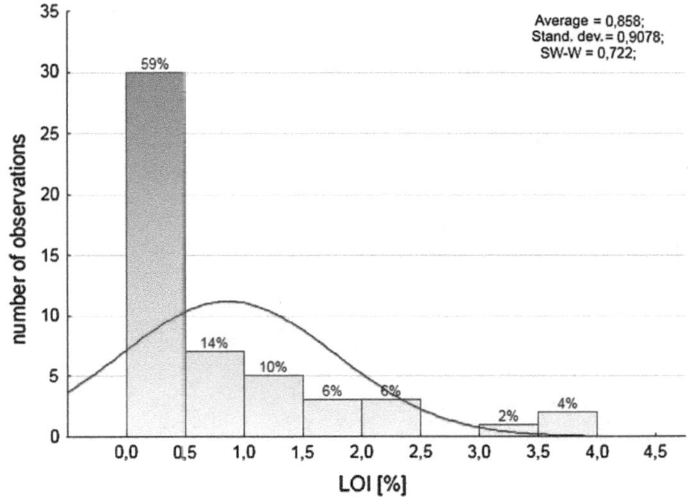

Fig. 4 Histogram of percentage share distribution of LOI [%] in the analysed samples of sediments from the Hel harbour

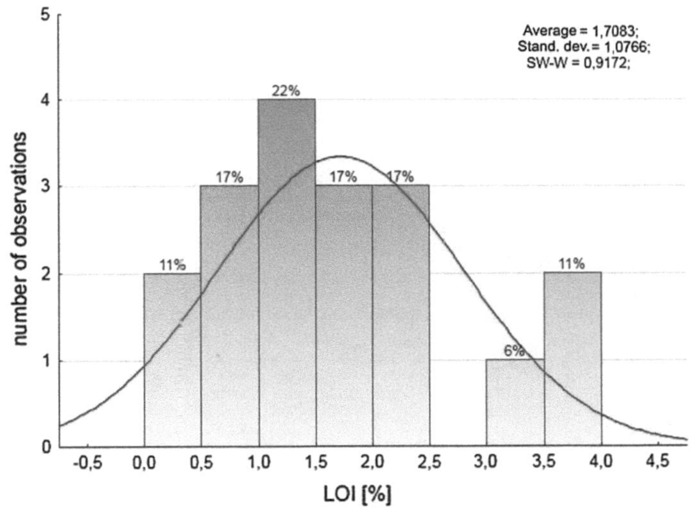

Fig. 5 Histogram of percentage share distribution of LOI [%] in the analysed samples of sediments from the inner part of the Hel harbour

and organic matter (Figs. 9 and 10). Correlation coefficient for the inner part, between <0.063 mm fraction material and magnetic susceptibility, with statistical significance $p = 0.05$, was $r = 0.57$ for fine-fraction particles and $r = 0.71$ for organic matter. Correlation coefficient values for the outer part of the harbour were

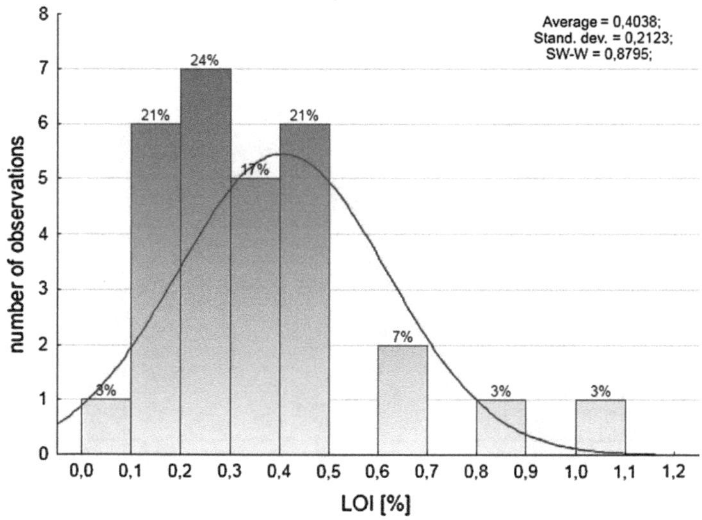

Fig. 6 Histogram of percentage share distribution of LOI [%] in the analysed samples of sediments from the outer part of the Hel harbour

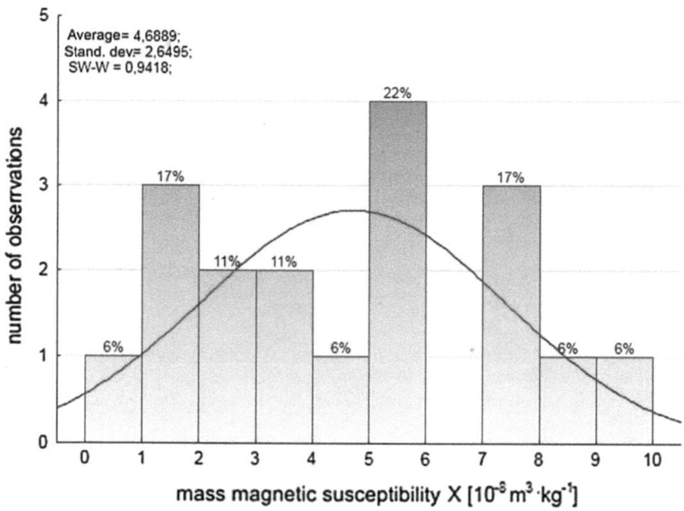

Fig. 7 Histogram of percentage share distribution of mass magnetic susceptibility in the analysed samples of sediments from the inner part of the Hel harbour

lower than in the inner part, with r = 0.21 and 0.07 for organic matter and fine-fraction particles, respectively. The granulometric content data make it possible to notice considerable differences between the inner and outer parts of the harbour. In the inner part, all the measured parameters were higher than in the outer

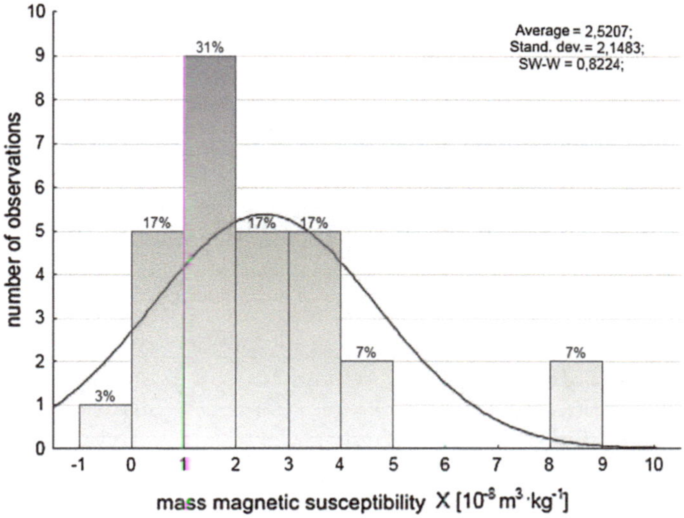

Fig. 8 Histogram of percentage share distribution of mass magnetic susceptibility in the analysed samples of sediments from the outer part of the Hel harbour

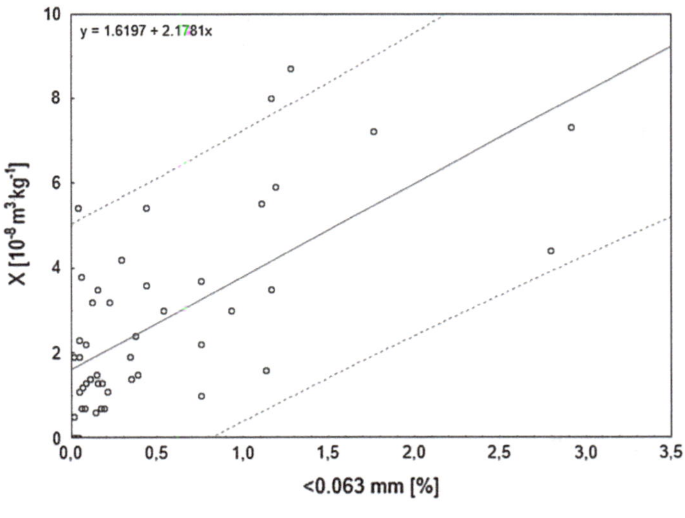

Fig. 9 Scatterplot of mass magnetic susceptibility (χ [10^{-8} m^3 kg^{-1}] and fine fraction [%])

part. This is caused mainly by limited water exchange and reduced water flow in the inner part. Stable hydrodynamic conditions characterising this part influence sediment sorting, physical processes affecting sediment content and depositional environment stability. In the case of the outer part, where the entrance to and the

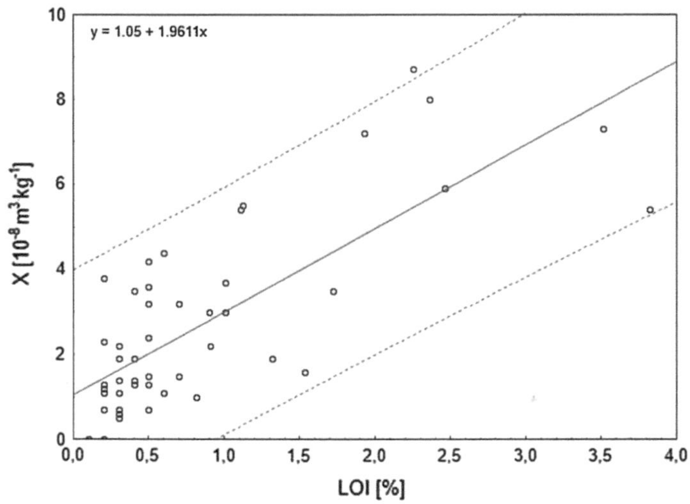

Fig. 10 Scatterplot of mass magnetic susceptibility (χ [10^{-8} m^3 kg^{-1}] and loss of ignition (LOI [%])

exit from the harbour are located and—therefore—direct water exchange between the harbour and the bay takes place, the measured parameters had lower values. The sedimentary material was well sorted, with low organic matter content and low magnetic susceptibility.

Magnetic susceptibility values measured in both the inner and outer part of the harbour are higher than those measured in the reference points. It is likely to be a result of pollutant accumulation in the sedimentary material within the harbour, caused by the nature of the depositional environment and the military function of the area.

4 Conclusions

The data of the physical parameters of the analysed sediments as well as of their magnetic susceptibility make it possible to conclude that the area of the naval harbour is subjected to considerable anthropopressure. At the same time, in the outer part of the harbour, the process of pollutant migration takes place, with pollutants being transported mainly further up the bay by water movement. An important aspect of the study was to present magnetic susceptibility as a significant parameter in the initial evaluation of marine sediment pollution. A comparison of magnetic susceptibility values for harbour sediments and reference points indicates a correlation between the degree of anthropopressure and the value of the measured parameter.

Moreover, the correlation between susceptibility and organic matter, and <0.063 fraction material may suggest that both of the above-mentioned parameters are a source of magnetically active particles, due to their sorption characteristics.

Increased sedimentary environment dynamics in the open part of the harbour result in better sorting of sediment particles, which causes a decrease in percentage share of <0.063 mm grain size fraction. It has also been observed that there exist statistically significant differences in the percentage share of <0.063 mm fraction between the inner part of the harbour (sampling points no. 1–10) and its outer part (no. 21–50). Close to the harbour entrance, the environment is more dynamic, as a result of which the rate of sedimentation in the outer part of the harbour is slower. The sediment is transported deep into the harbour. The largest deposition of sediments most probably takes place in the western section of the inner part of the harbour.

Magnetic methods used in the study may act as the first indicator in other environmental studies. Followed by a detailed chemical analysis, they will constitute a basis for the evaluation of the ecosystem under analysis, which in this particular case is the bottom sediment of the former naval harbour in Hel.

References

Blott SJ, Pye K (2001) GRADISTAT: A grain size distribution and statistics package for the analysis of unconsolidated sediments. Earth Surf Proc Land 26:1237–1248

Ciborowski T (2010) Substancja organiczna. In: Bolałek J. (ed.) Fizyczne, biologiczne i chemiczne badania morskich osadów dennych, Gdańsk: Wydawnictwo Uniwersytetu Gdańskiego, pp 287–290

Gołuchowska B, Strzyszcz Z, Kusza G (2012) Magnetic susceptibility and heavy metal content in dust from the lime plant and the cement plant in Opole Voivodeship. Archives of Environmental Protection 38(2):71–80

Krumbein WC (1964) Some remarks on the phi notation. J. Sedim. Petrol. 34:165–197

Łęczyński L, Szymczak E (2010) Własności fizyczne osadów dennych. In: Bolałek J (ed.) Fizyczne, biologiczne i chemiczne badania morskich osadów dennych. Gdańsk: Wydawnictwo UG, pp 63–115

Magiera T, Lis J, Nawrocki J, Strzyszcz Z (2002) Podatność magnetyczna gleb Polski. Atlas, PIG Warszawa

Myślińska E (2010) Laboratoryjne badania gruntów. Wydawnictwo Uniwersytetu Warszawskiego, Warszawa

Pasecki R (2010) Hel 1939–1945 Baza jednostek szkolnych i eksperymentalnych U-bootwaffe. Muzeum Obrony Wybrzeża. Stowarzyszenie "przyjaciele Helu". Zeszyt 10, p 64

Piekarek-Jankowska H, Wojcieszek D (2010) Metody pobierania próbek morskich osadów dennych. In: Bolałek J. (ed.) Fizyczne, biologiczne i chemiczne badania osadów dennych. Wydawnictwo Uniwersytetu Gdańskiego, Gdańsk, pp 19–23

Strzyszcz Z, Rachwał M, (2010) Zastosowanie magnetometrii do monitoringu i oceny ekologicznej technogennych pyłów oraz gleb na obszarach objętych wpływem emisji przemysłowych. Prace i Studia, No. 78, IPIŚ PAN, Zabrze 2010, pp 88

Udden JA (1914) The mechanical composition of plastic sediments. Bull. Geol. Soc. Am. 25: 655–744

Winters WJ, Dillon WP, Pecher IA, Mason DH (2000) "GHASTLI—determining physical properties of sediment containing natural and laboratory formed gas hydrate" Coastal Systems and Continental Margins—Natural Gas Hydrate in Oceanic and Permafrost Environments. Kluwer Academic Publishers, Dordrecht, Netherlands, pp 311–322

Wentworth CA (1922) Scale of grade and class terms for clastic sediments. J. Geol. 30:377–392

Zawadzki J, Fabijańczyk P, Magiera T, Strzyszcz Z (2010) Study of litter influence on magnetic susceptibility measurements of urban forest topsoils using the MS2D sensor. Environ. Earth Sci. 61(2):223–230

Zawadzki J, Magiera T, Fabijańczyk P, Kusza G (2012) Geostatistical 3-dimensional integration of measurements of soil magnetic susceptibility. Environ. Monit. Assess. 184(5):3267–3278

Magnetic Vertical Structure of Soil as a Result of Transformation of Iron Oxides During Pedogenesis. The Case Study of Soil Profiles from Slovakia and Ukraine

Maria Jeleńska, Beata Górka-Kostrubiec and Sylwia K. Dytłow

Abstract This work reports the study of depth differentiation of magnetic properties of four soil types: Kastanozem, Cambisol, Luvisol and Chernozem. The aim of the study is to find the way of magnetic minerals formation and transformation. The three profiles: Cambisol, Luvisol and Chernozem were taken from moderate temperature and moisture region from Slovak Republic and one profile, Kastanozem, was taken from semi-dry region of Southern Ukraine. Thermal behavior of saturation remanence revealed that magnetic composition of studied soils is a mixture of two components. Upper layers contain significant contribution of maghemite. This is demonstrated by unblocking temperature T_{ub} lower than 670 °C and higher than 600 °C and low values of coercivity B_c and B_{cr}. Variable amounts of hematite demonstrated by $T_{ub} \sim 670$ °C are seen for samples from parent rock horizons. Cumulative log Gaussian (CLG) analysis of the IRM acquisition curve revealed the presence of at least two components of coercivity along each profile. The variation of hematite content along profile was evaluated by depth variation of hard component of saturation remanence HIRM. The domain state of magnetic minerals derived from the Day's plot modified by Dunlop showed that all samples are grouped in the area between the mixing curve for single-domain SD and multi-domain MD grains and the mixing curve for SD and superparamagnetic SP grain. The depth variation of susceptibility χ, frequency dependence of susceptibility χ_{fd}, hysteresis parameters, anhysteretic remanence ARM, hard component of isothermal remanence HIRM, and relations between these magnetic parameters were used as indicators of development of pedogenesis. The results allow us to conclude that formation of iron oxides occurred by a transient intermediate phase—hydromaghemite which was transformed to hematite and/or to goethite depending on soil environment. The ratio χ_{fd}/HIRM allows to evaluate approximately similar precipitation for Kastanozem, Cambisol and Luvisol and

---

M. Jeleńska (✉) · B. Górka-Kostrubiec · S.K. Dytłow
Institute of Geophysics, Polish Academy of Sciences, Ks. Janusza 64, 01-452, Warsaw, Poland
e-mail: bogna@igf.edu.pl

M. Jeleńska et al. (eds.), *Magnetometry in Environmental Sciences*,
GeoPlanet: Earth and Planetary Sciences, DOI 10.1007/978-3-319-60213-4_8

much lower for Chernozem. As Chernozem should be formed in the same conditions as Cambisol and Luvisol, we probably observed the effect of higher permeability of parent rock in Chernozem profile sediments than in the case of Cambisol (granite) and Luvisol and Kastanozem (loess). Relation of loss on ignition LOI to magnetic parameters indicates that organic matter promotes pedogenesis progress. It is possible to relate vertical magnetic structure of soil to pedogenic differentiation into genetic horizons.

Keywords Environmental magnetism · Magnetic mineralogy and petrology · Rock magnetic properties

1 Introduction

Soil magnetometry (Evans and Heller 2003; Thompson and Oldfield 1986; Maher and Thompson 1999) uses magnetic properties of soil to study soil forming processes—pedogenesis (Orgeira et al. 2011; Maher et al. 2002, 2003; Jordanova and Jordanova 1999; Jordanova et al. 2010; Jeleńska et al. 2004, 2008a, b, 2010; Hanesch and Petersen 1999). Magnetic properties of soil depend on lithogenic properties of parent material and pedogenic processes occurring during soil formation. In particular, soil properties have been formed under the influence of five factors; (1) climate, (2) vegetation (fauna, flora and human activity), (3) topography, (4) parent material and (5) time (Fanning and Fanning 1989; Jenny 1941; Maher 1998; Maher et al. 2003; Orgeira et al. 2011).

Iron is a widespread element in the Earth's crust and fourth in abundance among all elements. In soils and parent material, Fe occurs among others in the form of Fe-oxides and hydro-oxides containing ferric and ferrous ions (Cornell and Schwertmann 2003). Concentration and redistribution of iron oxides and hydroxides indicate the intensity of pedogenic processes.

One of the first soil studies by magnetic methods were reported by Maher (1986). She investigated magnetic differentiation of several soil types with relation to parent material input, climate, fire and fermentation, atmospheric input, podsolization and gleying. Maher (1998) and Maher et al. (2003) made climatic reconstruction based on magnetic susceptibility of top soil in relation to such climatic parameters as temperature and humidity and such soil parameters as acidity, drainage and water balance. Studies were conducted on Chernozem from Russia and Ukraine and paleosoils from Chinese loess profiles (Liu et al. 2004, 2005; Maher and Thompson 1995; Maher et al. 1994). Hanesch and Schloger (2005) showed the influence of parent rocks on type and structure of magnetic particles and their distribution with depth. Jordanova et al. (2010) found for soils from Bulgaria that granulometry of magnetic minerals depends on textural properties of soil, humus and organic matter content and soil reactivity.

However, formation and transformation mechanisms of magnetic minerals during pedogenesis need further study. Complexity arises from the fact that magnetic properties may reflect conditions that prevailed under ancient climatic environment. Therefore, depth variations of soil magnetic structure developed in different environmental conditions are so important.

This work reports the study of depth differentiation of magnetic properties of four soil types: Kastanozem, Cambisol, Luvisol and Chernozem in order to find the way of magnetic minerals formation and transformation. The aim of the paper is to relate these processes to various environmental conditions, and to characterize on this basis individual horizons of different soil units.

2 Description of Soil Profiles

In 1998 the Union of Soil Sciences adopted the World Reference Base for Soil Resources (WRB) for soil relation and classification. WRB proposed 30 Soil Reference Groups gathered in 10 sets. For the study, four soil units belonging to 3 sets of Soil References Groups were chosen. Haplic Cambisol belongs to Set 5 which holds soils that are moderately developed due to their limited age or rejuvenation of soil material. Calcic Chernozem and Kastanozem belong to Set 8 which holds soils from arid and semi-arid regions. Haplic Luvisol belongs to Set 9 which holds brownish and grayish soils from humid temperate regions. The three profiles, Cambisol, Luvisol and Chernozem, were taken from moderate temperature and moisture region from Slovak Republic and one profile Kastanozem was taken from semi-dry region of Southern Ukraine. For the Slovak profile, the mean precipitation and mean temperature are about 450 mm/year and 9.8 °C; for the Ukrainian profile, the mean precipitation and mean temperature are 350 mm/year and 8.2 °C. The Slovak profiles: Cambisol, Luvisol and Chernozem were chose because they were formed under similar climatic conditions but they differ in parent rocks: granite, loess and carbonate fluvial sediments, respectively. The profiles belong to the soil categories which are fertile and mature with well developed humus layer and advance pedogenic processes. The Ukrainian profile, Kastanozem, was added for comparing the Slovak profiles with the soil of similar features but formed under different climate conditions.

Soil samples were collected along vertical profiles at depth intervals of 10 cm to reach the parent rock level from excavated soil pits. The profiles were from 100 to 170 cm long. The air dried and sieved (through a mesh with 1 mm openings) samples were used in measurement of physico-chemical properties of soil. The soil samples for magnetic measurements were stirred in water and formed into cylindrical specimens of a volume of about 10 cm^3 and mass of about 15 g. Smaller specimens, of volume of about 0.5 cm^3 and mass of 0.5 g, were prepared for thermal and magnetic hysteresis measurements. The measurements were made for 3–4 specimens from each 10 cm soil layer (0–10 cm, 10–20 cm, etc.). The magnetic and soil parameters were computed from the measurements repeated several times. The measurement errors of magnetic and soil parameters were no more than 1 and 5%, respectively.

3 Methods

3.1 Measurement of Magnetic Parameters

Magnetic characteristics of soil were obtained by a number of measurements of magnetic parameters.

Low-field magnetic susceptibility was measured by use of the Multi-Function Kappabridge MFK1-FA (AGICO, Czech Republic). The susceptibility value was normalized by soil mass to get mass specific susceptibility χ and expressed in 10^{-8} m^3 kg^{-1}.

The same device was used for determination of frequency dependence of susceptibility (χ_{fd}) which defines decrease of susceptibility with increasing frequency $\chi_{fd} = \chi_{lf} - \chi_{hf}$, where χ_{Lf} is the susceptibility measured in 976 Hz magnetic field and χ_{Hf} is the susceptibility measured in 15,600 Hz field. Then the values were recalculated to the frequency ratio 1:10 usually used in references according to the formula given by Hrouda (2011). The percentage value of $\chi_{fd\%}$ is expressed as $((\chi_{lf} - \chi_{hf})/\chi_{Lf})\cdot 100\%$.

Anhysteretic remanent magnetization ARM was acquired at 100 mT of AF field and 100 μT of DC field using the device LDA-3 made by the AGICO. ARM was measured by the SQUID magnetometer (2G Enterprises, USA). The value of anhysteretic susceptibility χ_{ARM} was calculated by dividing the value of ARM by the value of DC magnetic field (100 μT).

The measurements of hysteresis loops were performed on small samples of 0.5 cm^3 using the Vibrating Sample Magnetometer (VSM) produced by the Molspin (Great Britain), operating at maximum magnetic field of 1T. The following parameters were determined from the hysteresis loop: saturation magnetization (M_s), saturation remanence (M_{rs}, SIRM) and coercivity B_c. The curve of subsequent DC back-field remagnetization of M_{rs} was measured on the same sample and used to determine the coercivity of remanence B_{cr}.

The curves of continuous thermal demagnetization of saturation isothermal remanent magnetization SIRM(T) were made using the device built by the Electronics (TUS), Warsaw, Poland. Before measurement, samples were magnetized in a field of 9 T, high enough to saturate almost all hard magnetic minerals. Thermal decay of SIRM was measured in magnetic screen during heating in air to 700 °C and cooling to room temperature. During thermal demagnetization, the components of remanence associated with the magnetic minerals were identified by their unblocking temperature (T_{ub}) values (Kądziałko-Hofmokl and Kruczyk 1976).

Magnetic susceptibility (χ), saturation magnetization (M_s) and saturation remanence (M_{rs}, SIRM) are the parameters which depend on concentration of magnetic minerals in soil. The parameter χ_{fd} indicates the presence of ultra-fine grains lying at the boundary of superparamagnetic (SP) and single domain (SD) size (Dearing et al. 1996). Percentage value of χ_{fd} ($\chi_{fd\%}$) is controlled by grain size distribution (GSD) of SP and SD grains and is used as a proxy of relative content of pedogenic SP + SD and PSD (pseudo-single domain) particles (Eyre 1997;

Worm 1998). Variations of this parameter with depth allow to determine the depth to which pedogenic processes take place. Anhysteretic susceptibility (χ_{ARM}) which has the greatest value for single domain (SD) magnetic grains is used as an indicator of SD grains content (Liu et al. 2004). The ratios of hysteresis parameters: M_{rs}/M_{rs} and B_{cr}/B_c presented on Day plot (Day et al. 1977) describe domain structure of magnetic grains. HIRM = 0.5·(SIRM + $IRM_{-0.3T}$), where SIRM is saturation isothermal remanence obtained in 1T field and $IRM_{-0.3T}$ is remanence remained after back-field of 0.3T treatment (Liu et al. 2013). HIRM is used as an indicator of hard minerals content. As in soil we have usually a mixture of at least two magnetic fractions, magnetic coercivity components were isolated based on cumulative log Gaussian (CLG) analysis of the IRM acquisition curve using IRMUNMIX program by Kruiver et al. (2001). IRM acquisition curves were obtained by applying irregularly spaced steps of magnetic field up to 3.4 T.

3.2 Measurement of Soil Parameters

The following soil parameters were measured: reactivity (pH), the approximate amount of organic matter obtained by applying the loss on ignition (LOI%) method, and the content of carbonates ($CaCO_3$). Laboratory analyses were carried out in accordance with the procedures of Polish Standard Methods.

Measurements of pH level were made by applying electrometric method using a digital meter pH/mV (ORP) Greisinger GPHR 1400. The samples were prepared according to instruction of Head (1992). Soil was sieved through 2 mm grid and the same mass of soil was diluted in distilled water in 1:5 soil-H_2O proportion and left. pH was measured after 24 h.

The loss on ignition (LOI) method is based on measuring the loss of sample mass after ignition of organic matter. The soil samples of known mass were placed in a ceramic vessel and heated at 550 °C for 2 h, then cooled in a desiccator and re-weighed. Organic matter content is expressed as a percentage difference between the initial and final sample mass divided by the initial sample mass.

Loss on ignition (LOI) has been widely used as a method to estimate the amount of organic matter (and indirectly of organic and inorganic carbon) in sediments because of its simplicity and facility of performance in such cases where more accurate Tiurin method is not necessary (Heiri et al. 2001; Santisteban et al. 2004). According to Hoogsteen et al. (2015), the method can be used under standard protocol applied.

Determination of the carbonate content was made by Scheibler`s volumetric method (Canarache et al. 2006). Carbonates contained in the sample were converted into CO_2 by adding 10% hydrochloric acid to the soil. The pressure of the released CO_2 raises level of water in a burette. The carbonate content (interpreted as an equivalent of calcium carbonate concentration) can be calculated from the difference in level of water in burette.

4 Description of Chemical and Magnetic Results

4.1 Chemical Results

The chemical characteristics (Fig. 1) are the most regular for Luvisol SEN and Chernozem SNO with increase of pH from 5.0–7.5 to 8.0–9.0 and increase of $CaCO_3$ content from 0 to 30–20%. $CaCO_3$ content shows similar depth variation as pH. Loss on ignition LOI decreases in both profiles from 8–9% to 2% in the bottom.

In Cambisol RUS and Kastanozem ASK, changes of pH and $CaCO_3$ content are not regular with increase in the middle part of the profiles and fall down in the bottom. $CaCO_3$ content for RUS in the ochric and cambic horizons is about 8–10% of carbonates, the bottom part of profile is enriched with more than 20% of $CaCO_3$. Organic matter content in both profiles, RUS and ASK fluctuates in the range of 4–8%, and suddenly falls to about 3–4% in the parent horizon (Fig. 1).

4.2 Magnetic Results

Identification of magnetic mineralogy were made on the basis of decay curves of SIRM during heating and by means of hysteresis parameters. SIRM(T) curves (Fig. 2) show that magnetic composition of studied soils is a mixture of two components. Upper layers contain significant contribution of maghemite. This is demonstrated by unblocking temperature T_{ub} lower than 670 °C and higher than 600 °C and low values of B_c and B_{cr} (Fig. 3). In the uppermost layer of Cambisol profile RUS, magnetite is present. Variable amounts of hematite demonstrated by $T_{ub} \sim 670$ °C are seen for samples from parent rock horizons. The greatest amount of hematite is observed in calcic Chernozem SNO in C and D horizons. The variation of hematite contribution along profile can be evaluated by depth variation of HIRM. Liu et al. (2013) showed excellent proportionality between hematite content and HIRM value in modern soil profiles from China. Contribution of another hard mineral—goethite to HIRM was excluded because goethite has a very low IRM value. Depth variations of HIRM (Fig. 3) show that absolute value of hematite content increases towards the surface of the profile; however, the percentage contribution of hematite, on the contrary, increases with depth towards the bottom and becomes significant in the parent rock horizons. The increasing percentage contribution of hematite with depth is also demonstrated by an increase of B_c and B_{cr} values (Fig. 3) in deeper parts of profiles. IRM acquisition curves were measured for two samples from each profile: for sample from the uppermost layer (A horizon) and for bottom sample (parent rock horizon). CLG analysis revealed the presence of at least two components of coercivity (Table 2; Figs. 4 and 5) in each sample. Three components were found in bottom sample from ASK profile and for both samples from SEN profile. The first components are usually very soft

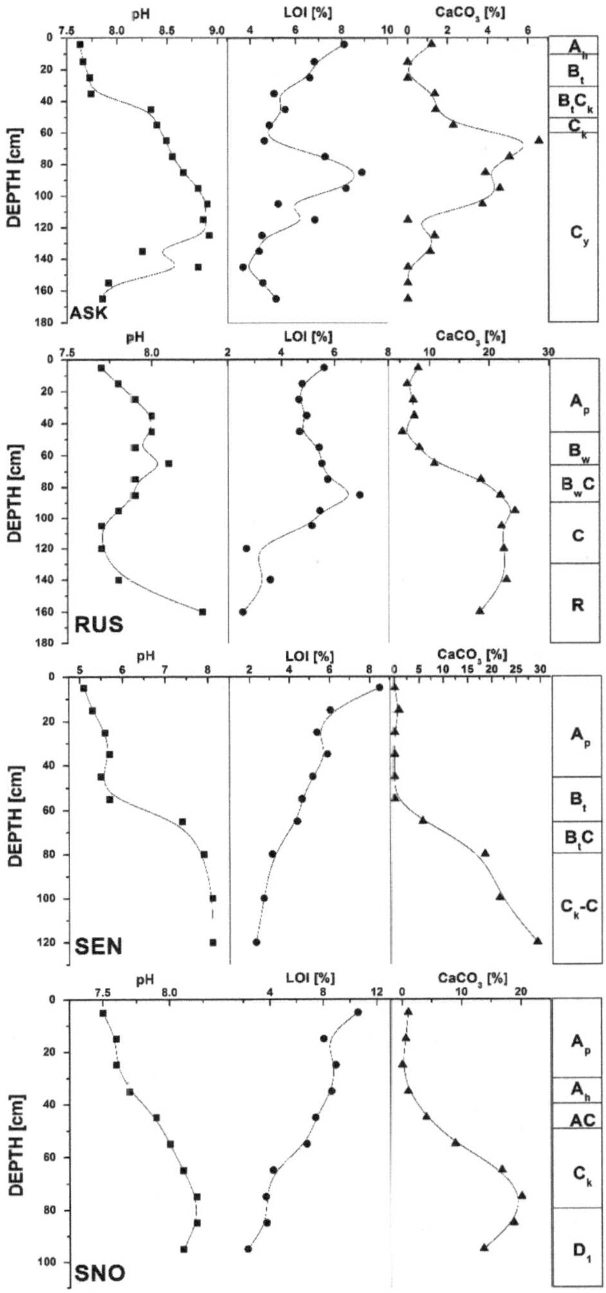

Fig. 1 The chemical characteristic of: **a** Haplic Luvisol (SEN); **b** Haplic Cambisol (RUS); **c** Kastanozem (ASK) and **d** Calcic Chernozem (SNO). First curve—reactivity pH; second curve—loss on ignition LOI and third curve—carbonates $CaCO_3$ content. Additionally, horizon distribution along depth is presented (Table 1)

Table 1 Description of the profiles

Sampled profile genetic horizons sequence	Description
Kastanozem—ASK Askania Nove, Ukrainian Steppe	Developed on loess under the short grass of steppe belt
A_h-B_t-B_t/C_k-C_k-C_y-C	A_h mollic horizon reach in humus, B_t argic brown horizon, C horizon with accumulation of lime and gypsum
Haplic Cambisol—RUS Rusovce, Slovak Republic	Developed on the stony weathering products of non-carbonate rocks—granite
A_p-B_w-B_w/C-C-R	A_p ochric horizon over a B_w cambic horizon B_w/C transitional horizon located over a parent material R mother rock—granite
Haplic Luvisol—SEN **Senec, Slovak Republic**	Formed on loess enriched in calcium carbonates
A_p-B_t-B_t/C-C_k-C	A_p ochric grayish horizon, poor in organic matter compared and containing less than 0.6% of carbonates B_t argic illuviation horizon with accumulation of clay B_tC transitional horizon C_k calcic and C horizons occurring below the reddish brown argic horizon Profile deprived of E-eluviation horizon
Calcic Chernozem-SNO **Stvartocna Ostrove,** **Slovak Republic**	Formed on old carbonate fluvial sediments
A_p-A_h-AC-C_k-D_l	A humus mollic horizon AC transitional horizon C_k calcic horizon D_l mineral base

and appear in small amount (Table 2). The most prevailing component is moderately hard fraction of $B_{1/2}$, ranging from 44 to 66 mT. $B_{1/2}$ is the applied field at which the phase acquires half of its saturation remanence value, providing a measure of the mean coercivity of the phase. The third component is very hard, of $B_{1/2}$ more than 100 mT. In SEN profile, this component reaches about 45%. Very soft component can most likely reflect contribution of fine grains on the boundary of SP/SD size. The main components, with $B_{1/2}$ of 44–66 mT, have a broad range of coercivity distribution with DP (DP is the dispersion parameter) of more than 0.4, corresponding to $B_{1/2}$ between 14 and 290 mT. It can indicate that this population consists of a wide distribution of grain size from MD to SD. Third component can represent lithogenic hematite. Samples from the bottom and the uppermost layer both consist of two or three components but always the coercivity of the upper sample is lower than the coercivity of the bottom sample (Fig. 5). For RUS and SEN profiles, these differences in $B_{1/2}$ values for the top and bottom samples are significant, for SNO profiles there are hardly seen.

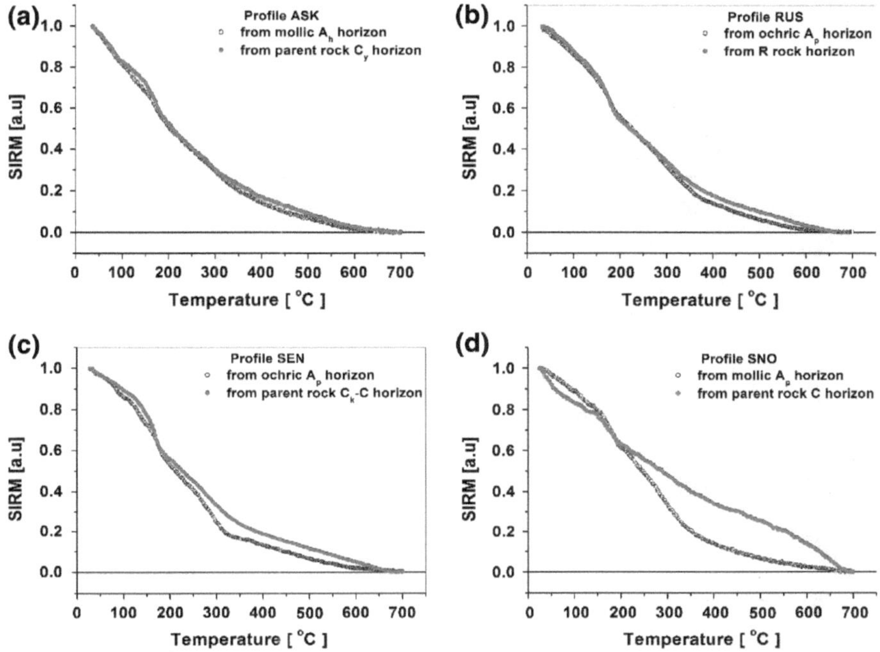

Fig. 2 Examples of SIRM(T) decay curves for *top* and *bottom* samples of profiles: ASK (**a**), RUS (**b**), SEN (**c**) and SNO (**d**)

The domain state of magnetic minerals has been derived from Day's plot (Day et al. 1977) modified by Dunlop (2002a, b) which presents M_{rs}/M_s versus B_{cr}/B_c relationship (Fig. 6). All the samples from shallow horizons, A and B, except SNO, are grouped in the area between the mixing curve 3 for SD and MD grains and the mixing curve for SD + 10 nm SP grain. Such position can be the result of influence of PSD or SP grains. For RUS, SEN and SNO profiles, samples taken from deep horizons (C, D_1, R) are scattered around the area occupied by samples taken from shallow horizons. For ASK, hysteresis parameters show the best grouping. For RUS, the values of H_{cr}/H_c are shifted towards high values and M_{rs}/M_s ratios are shifted towards low values. This can be a result of relatively low content of hematite in RUS profile and may be due to interaction of SD maghemite grains.

Magnetic structure of soil which reflects differences in genetic characteristics of horizons usually are analysed by depth variation of magnetic parameters and relations between them. Variation of magnetic susceptibility with depth shows several common features (Fig. 7). For all soil units, maximum values of χ occur in upper layers and slowly decease along profile to minimum values in parent rocks. Besides this common pattern, significant differences are observed. The highest maximum is seen in ochric Ap horizon (107×10^{-8} m^3/kg at 25 cm) for Luvisol profile SEN. The smaller and shallower maximum occurs in Kastanozem ASK on

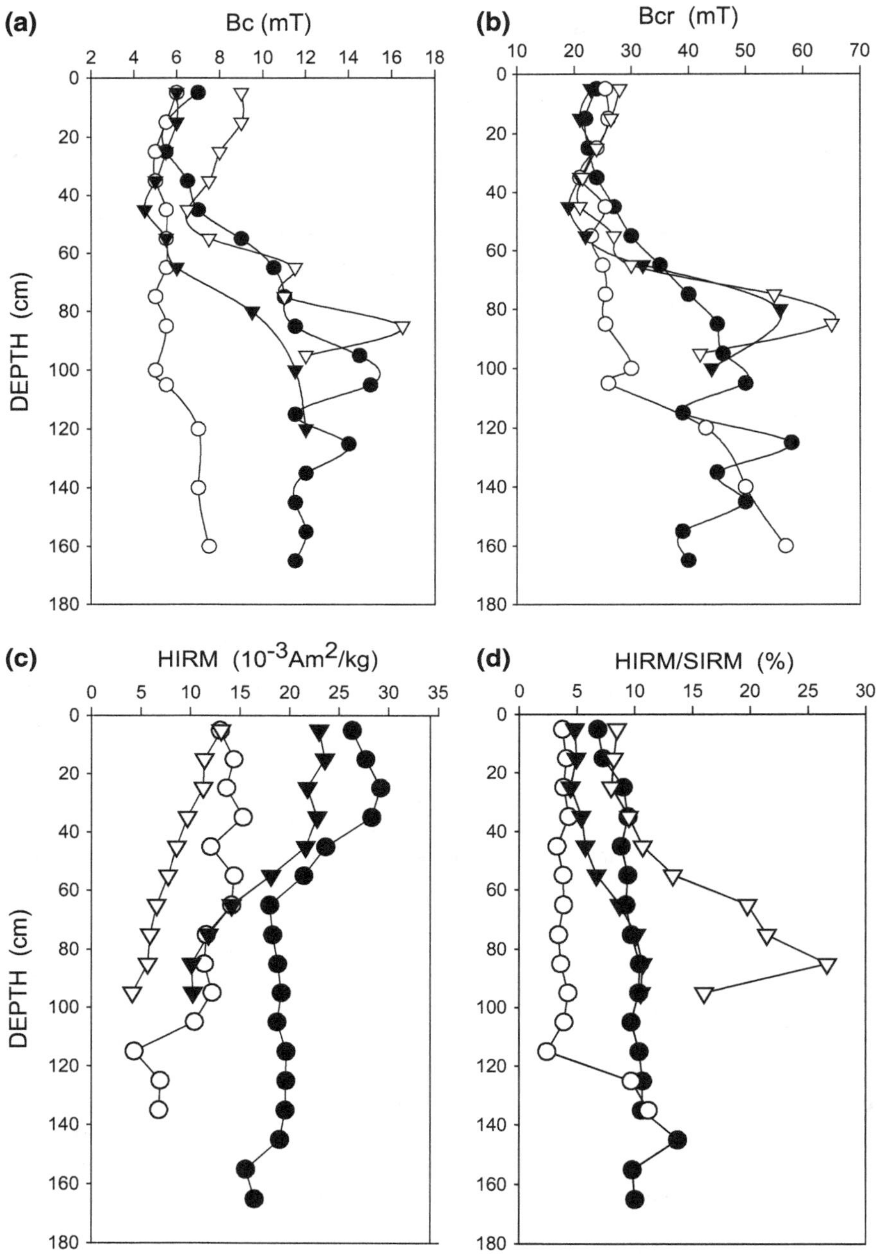

Fig. 3 Depth variations of B_c (**a**), B_{cr} (**b**), HIRM absolute values (**c**), and HIRM values related to SIRM (**d**). *Full circles* ASK profile; *open circles* RUS profile; *full triangles* SEN profile; *open triangles* SNO profile

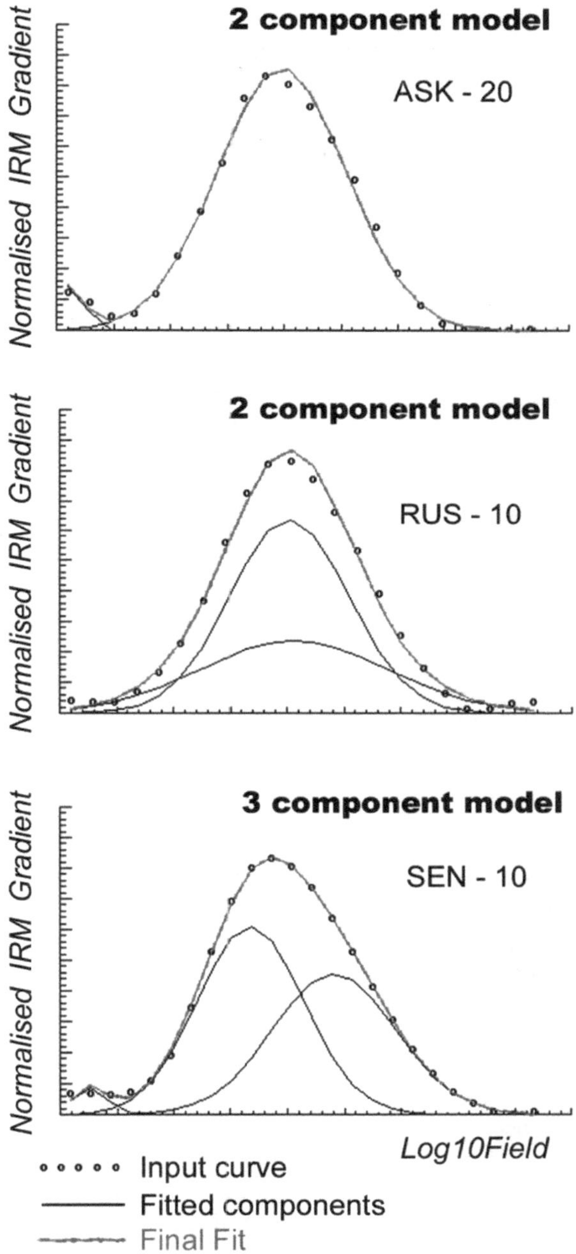

Fig. 4 Examples of GLC analysis of IRM acquisition curves

Table 2 Cumulative analysis of SIRM components

Sample	Comp	SIRM (%)	$B_{1/2}$ (log mT)	$B_{1/2}$ (mT)	DP (log mT)	Range (mT)	B_c (mT)	B_{cr} (mT)
ASK 20 cm	1	3.07	0.15	1.41	0.045	1.3–1.6	5.5	22
	2	96.9	1.645	44.2	0.471	15–131		
ASK 150 cm	1	6.16	0.199	1.58	0.119	1.2–2.1	12	40
	2	73.3	1.754	56.7	0.424	21–151		
	3	20.5	2.209	161.8	0.246	92–285		
RUS 10 cm	1	63.3	1.683	48.2	0.435	18–131	6	25.5
	2	36.7	1.718	52.2	0.689	11–255		
RUS 130 cm	1	6.04	0.219	1.65	0.136	1.2–2.3	7	45
	2	93.9	1.822	66.3	0.447	24–186		
SEN 10 cm	1	2.35	0.217	1.65	0.126	1.2–2.2	6	22
	2	52.4	1.399	25.1	0.389	10–61		
	3	45.2	2.013	103.0	0.449	37–290		
SEN 100 cm	1	7.23	0.078	1.20	0.173	0–1.8	11.5	44
	2	48.4	1.782	60.5	0.469	20–178		
	3	44.4	2.178	150.8	0.483	49–458		
SNO 10 cm	1	4.8	0.195	1.57	0.113	1.2–2	9	27
	2	95.2	1.74	54.9	0.455	19–157		
SNO 100 cm	1	12.4	0.384	2.4	0.23	1.4–4.1	12	42
	2	87.6	1.768	58.6	0.59	15–228		

SIRM relative saturation remanence of component; $B_{1/2}$ the applied field at which the mineral phase acquires half of its saturation IRM; *DP* the dispersion parameter of the coercivity, B_c coercivity, B_{cr} coercivity of remanence

Ah/Bt boundary (81×10^{-8} m^3/kg, at 15 cm). Hardly seen maximum is observed in horizon Br (90×10^{-8} m^3/kg, at 55 cm) in Cambisol profile RUS, and no maximum is found in calcic Chernozem SNO (24×10^{-8} m^3/kg). Calcic Chernozem exhibits very low χ values along whole profile, whereas Haplic Cambisol is characterized by high values of χ observed to extremely deep layers below 120 cm.

Anhysteretic susceptibility χ_{ARM} follows the changes of χ with one exception—for Kastanozem, the values of χ_{ARM} remain high even for Cy and C horizons.

Variations of $\chi_{fd\%}$ demonstrate similar changes with depth as χ and χ_{ARM}. $\chi_{fd\%}$ reaches high values between 10 and 12% for upper horizons A and B, then diminish to 0–2% for parent rock with exception of Kastanozem for which $\chi_{fd\%}$ never falls below 8%. For Calcic Chernozem, $\chi_{fd\%}$ is low, of about 6% for A and A/C horizons, whereas in Cambisol profile, $\chi_{fd\%}$ exhibits almost constant values of about 10% down to deep layers below 100 cm.

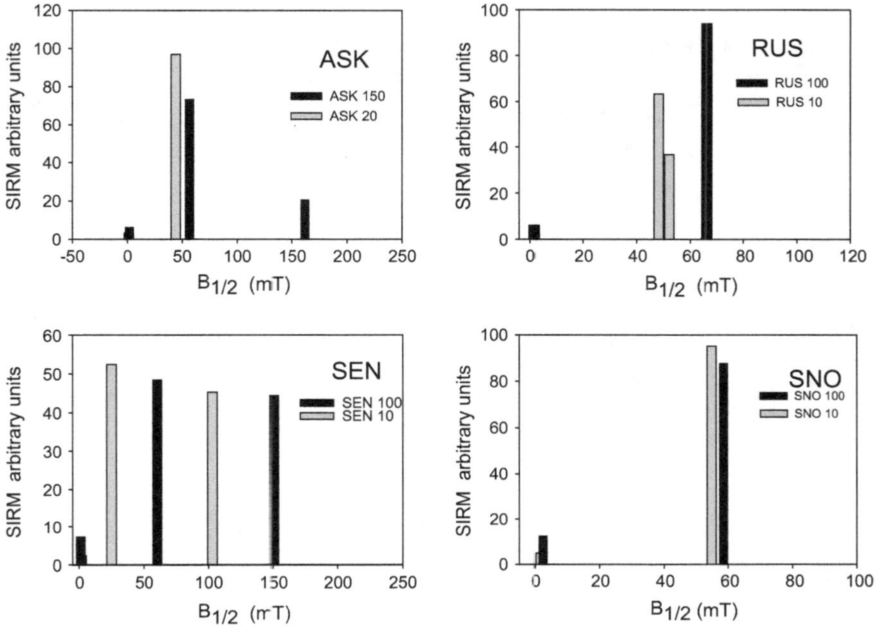

Fig. 5 Distribution of coercivity components calculated by CLG analysis of IRM acquisition curves for top and bottom layers of studied profiles. SIRM is the intensity of component related to total intensity of remanence; $B_{1/2}$ is the value of the applied field at which the phase acquire half of its saturation remanence

Fig. 6 Day's plot (Day et al. 1977) modified by Dunlop (2002a, b). See notation in Fig. 3

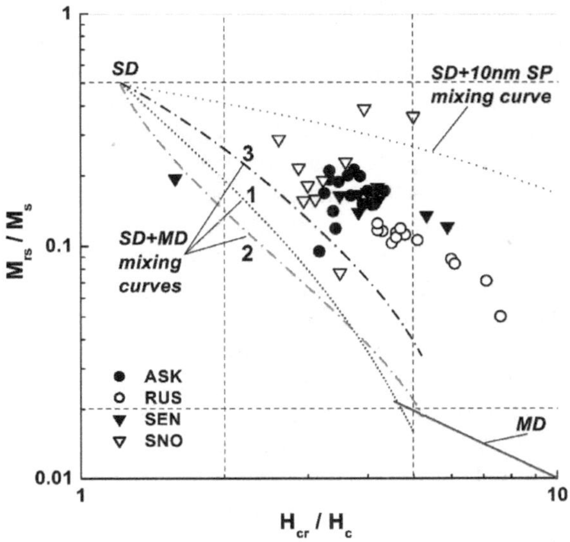

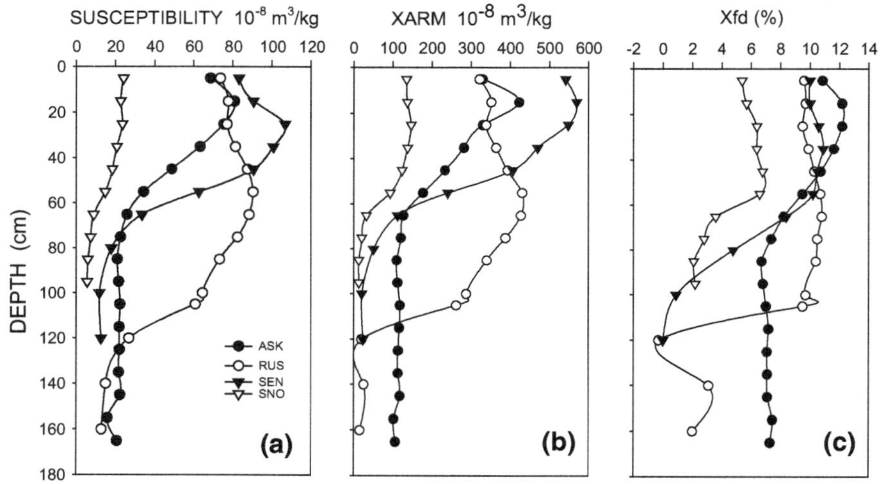

Fig. 7 Vertical distribution of magnetic parameters: magnetic susceptibility χ (**a**), anhysteretic susceptibility χ_{ARM} (**b**) and frequency dependence of magnetic susceptibility $\chi_{fd\%}$ (**c**). See notation in Fig. 3

5 Discussion

The magnetic enhancement defined as $ME = \chi - \chi_{bot}$ where χ_{bot} is the value of χ for parent rock, expresses an increase of χ observed in upper horizons. ME is related to the increase of the amount of strong ferrimagnetic minerals maghemite and/or magnetite of nano-sized SP grains which were produced during pedogenesis. There are two hypothesis of maghemite formation in soil. The first hypothesis assumes that nano-sized magnetite is formed directly by reduction of Fe^{3+} to Fe^{2+} in the presence of iron-reducing bacteria and organic matter (Mullins 1977; Van Velzen and Dekkers 1999). Magnetite is then oxidized to maghemite. Another alternative hypothesis proposed by Barrón and Torrent (2002) and Torrent et al. (2010) assumes formation of a transient intermediate phase—hydromaghemite from ferrihydrite which, in turn, is a product of weathering of Fe-bearing minerals such as smectite (Schwertmann 1985, 1988). Hydromaghemite is then transformed to hematite and/or to goethite depending on soil environment. Liu et al. (2008) showed that the grain size of hydromaghemite increases on aging from SP to SD range before transformation to hematite. The authors postulated that grain size distribution (GSD) of maghemite depends on the rate of the following processes: weathering of Fe-bearing minerals which creates ferrihydrite, transformation of ferrihydrite into hydromaghemite, growth of hydromaghemite grains from SP to SD size, and transformation of hydromaghemite into hematite. In moderately weathered soils one should expect continuous formation of maghemite in all intermediate stages and wide GSD, and presence of hematite. Such pathway of transformation of ferrihydrite is the reason of heterogeneity of composition of magnetic minerals and is controlled by environmental factors such as precipitation and temperature.

5.1 Iron Forms in the Studied Soil Profiles

SIRM(T) curves indicate that maghemite is the main ferrimagnetic mineral. As it is shown by soft IRM versus hard IRM (HIRM) diagrams (Fig. 8), maghemite was formed together with antiferromagnetic hematite. The relations are linear for all profiles but the regression lines differ in the slope and regression coefficient R2, being the weakest for ASK 0.66 and the strongest for SEN 0.95. In general, this relationship indicates formation of both minerals during a common process. Maghemite content is strongly related to pedogenic processes as is seen from soft IRM versus χ (R2 = 0.92) and versus χ_{fd} (R2 = 0.89) relationships (Fig. 9a and b). Only SNO data declined from the common line. Hematite creation expressed by HIRM also shows relation to pedogenic processes seen in HIRM versus χ and χ_{fd} diagrams (Fig. 9c and d). In this case, relations, although linear, are different for each profile, being the weakest for ASK and the strongest for SEN as in case of soft to hard minerals relations. Different slopes for HIRM versus χ and χ_{ARM} coefficient for profiles indicated differences in pedogenesis rate.

CLG analysis of IRM acquisition curves revealed that two or three components of coercivity are present along all profiles. Bottom of the profiles contains harder minerals than surface layers. The variations of HIRM and soft IRM with depth (Fig. 3) indicate that the influence of lithogenic hematite was significant from the

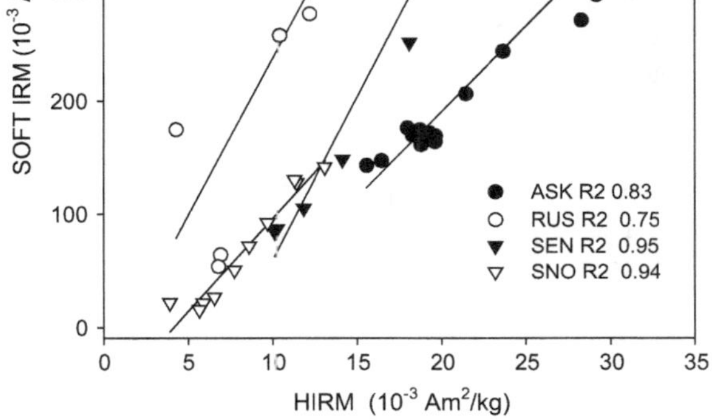

Fig. 8 Soft component of IRM (SOFT IRM) versus hard component of IRM (HIRM) for studied profiles. R2 values for linear relationship are shown in the lower right corner of diagram. See notation in Fig. 3

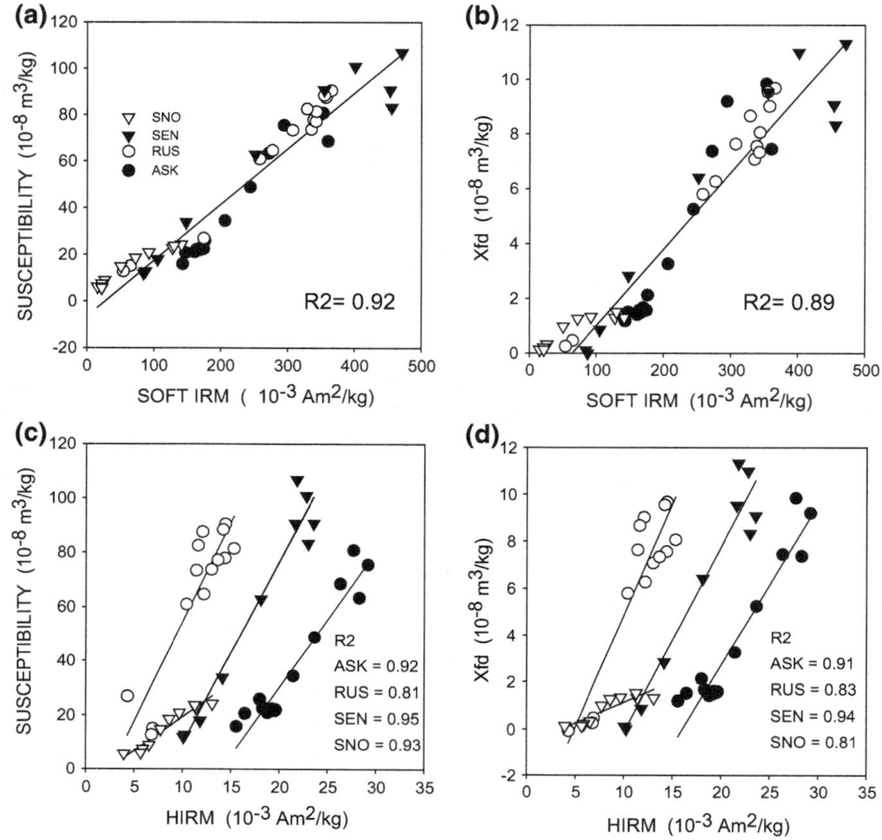

Fig. 9 Soft component of IRM (SOFT IRM) versus susceptibility (**a**) and versus χ_{fd} (**b**). Hard component of IRM (HIRM) versus susceptibility (**c**) and versus χ_{fd} (**d**). See notation in Fig. 3

bottom to the depth of about 60 cm (transitional horizons). The soft magnetic minerals, maghemite and magnetite, produced during pedogenesis are important in genetic horizons A and B.

Magnetic enhancement ME is strongly related to χ_{fd} which is used to determine content of fine magnetic particles in small grain size range on SP/SD boundary (Fig. 10a). Similar relation exists between χ_{fd} and χ_{ARM} which is used as a proxy for SD grains concentration (Fig. 10b). In both plots, the data shows a common regression line indicating that the studied profiles have similar grain size distribution. The results presented support rather the second pathway of magnetic minerals formation.

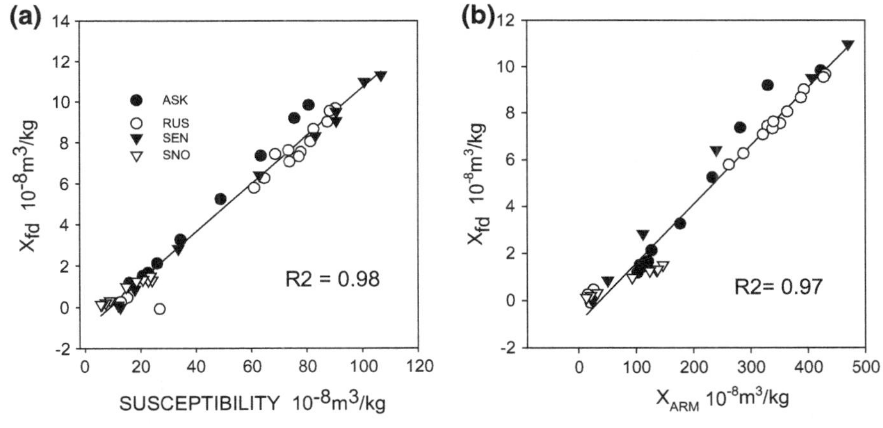

Fig. 10 Magnetic susceptibility χ (**a**) and anhysteretic susceptibility χ_{ARM} (**b**) in relation to χ_{fd}

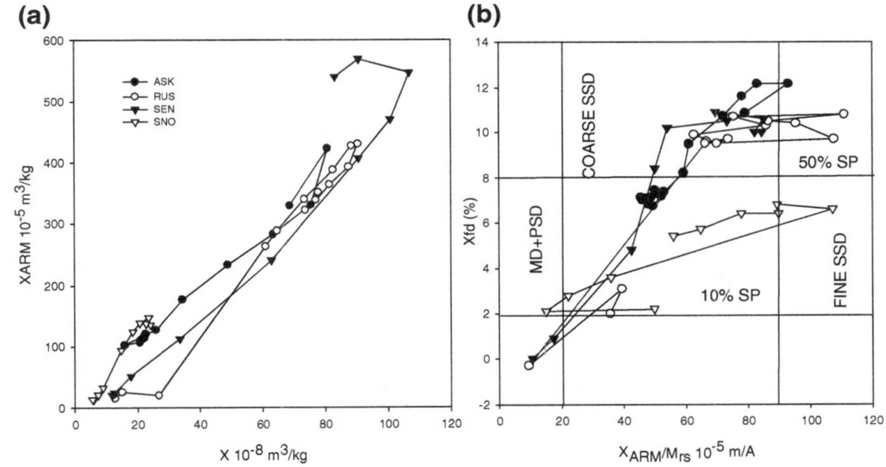

Fig. 11 χ_{ARM} versus χ (**a**), and $\chi_{fd\%}$ versus χ_{ARM}/M_{rs} (**b**)

5.2 *Magnetic Characteristics of Genetic Horizons*

In previous section we considered the profiles as a whole. Now we shall discuss magnetic structure of particular horizons for particular profiles. King et al. (1982) proposed a rapid method for evaluation of the relative grain size of magnetite or maghemite based on the graph of χ_{ARM} versus χ. Such graphs for studied profiles show that, in general, in the middle part of all profiles, χ_{ARM} versus χ can be approximated by linear relation. The constant value of χ/χ_{ARM} ratio (Fig. 11a) observed in the middle part of the profiles means that the magnetic minerals were produced from the bottom to the top with the same GSD. However, in the upper layers in A horizons, significant differences between soil units are observed. Also parent rock horizons show different GSD.

As $\chi_{fd\%}$ depends mainly on GSD around SP/SD boundary and is reversely related to the width of GSD in this range, $\chi_{fd\%}$ is a good demonstration of progress in pedogenesis. Diagram of $\chi_{fd\%}$ versus χ_{ARM}/M_{rs} proposed by the Dearing et al. (1996) allows to determine the progress in pedogenesis in relation to domain state (Fig. 11b). Domain structure and GSD seen in the Dearing plot show that the largest grains of MD and PDS structure with about 10% of SP grains contribution are in the parent rock horizons. In B horizon, SP grains content increases to 50% and grain size decreases towards coarse SSD. In A horizons, SP grains content increases to 70% and grain size changes from coarse SSD to fine SSD structure. There are two exceptions: ASK profile starts from about 6–7% of SP content and small grain size of SSD structure in parent rock horizon, and SNO profile contains only 5% of SP grains in A horizon. The finest grains are found in B horizon of the SNO profile.

Soil pH is regarded as a very important factor which influences pedogenic processes occurring in soil. However, we do not observe any correlations between pH and magnetic parameters. We observed some influence of LOI on magnetic properties of soil (Fig. 12). For SEN and SNO profiles, soft and hard components of IRM increase linearly with increase of LOI. For RUS the relation is hardly seen. For ASK the creation of magnetic minerals is independent of LOI.

In addition, χ and χ_{ARM} show a linear trend with an increase of LOI (Fig. 12). It suggests that organic matter promotes development of magnetic structure characteristic for advanced pedogenic processes with creation of fine grains of strong magnetic minerals. On the other hand, too high content of organic matter hampers pedogenic processes. This is demonstrated by decreasing all magnetic parameters in layers with highest LOI values. It is well seen on LOI versus χ and χ_{ARM} plots (Fig. 12c and d).

Now it is possible to relate magnetic structure to pedogenic characteristics of studied profiles. The Luvisol is characterized by a marked textural differentiation within the profile. The surface horizon is depleted of clay which accumulates in the subsurface horizon. The translocation of clay from the surface profile to the depth of accumulation leads to formation of illuviation (*argic*) horizon.

The surface soil may become depleted of clay and free of iron oxides to the extent that a greyish eluviation B_t horizon forms under a dark but thin A-horizon (Driessen et al. 2001). The process of mobilization of clay in the mollic horizon and their transport to the accumulation illuvic (B_t) horizon where the clay is accumulated, is characteristic of this type of soil and manifests itself in high χ and SP content in B_t and stabilization of these parameters in A horizon.

Cambisol is a relatively young soil created in short time. The distinctive feature of Cambisols is the beginning of horizon differentiation which developed in a transitional stage from a young soil to a mature soil. The first step of Cambisol development is the formation of a cambic subsurface horizon. The cambic horizon B_w is situated between an A-horizon and a relatively unaltered C-horizon. The process of formation of a cambic horizon is fundamentally the same in all climate

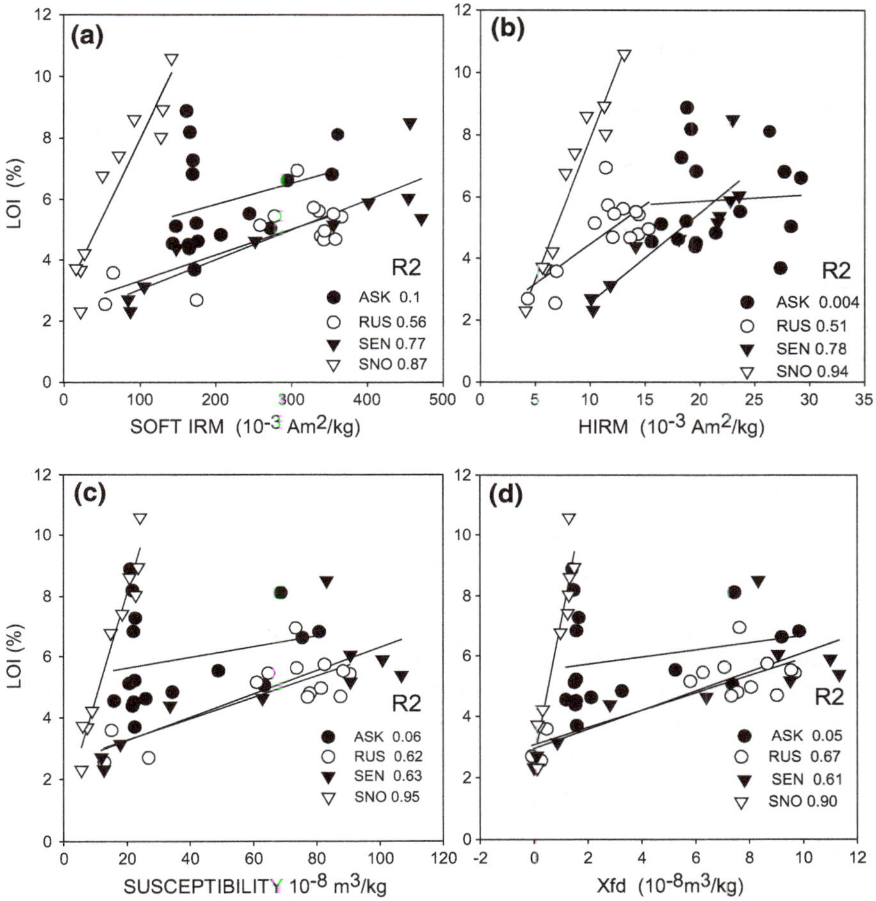

Fig. 12 Relation of loss on ignition (LOI) to magnetic parameters: LOI versus SOFT IRM (**a**), LOI versus HIRM (**b**), LOI versus χ (**c**) and LOI versus χ_{fd} (**d**). See notation in Fig. 3

zones but the intensities of chemical and biological transformations depend on climate conditions and may take up shorter or longer periods. RUS profile has been formed in wet climate with relatively long lasting warm periods characteristic of southern Slovakia. Such climatic conditions promoted quick washing down of calcium carbonates and disintegration of primary minerals into secondary clay minerals. In alkaline conditions (pH = 7.75–8.25) iron oxides and hydroxides are not mobile and together with secondary minerals gradually gather in the upper part of profile. This prevents farther pedogenesis and fixed magnetic structure of soil. This process results in rapid changes of all magnetic parameters between horizons lower than cambic layer and higher than cambic layer. From 100 cm upwards, magnetic parameters of RUS profile remain almost unchanged.

The Calcic Chernozem formed under well drained conditions and with the approximate equality between the annual precipitation sum and evaporation. In this type of Chernozem, the annual precipitation sum is about 450 mm. Profile SNO was formed on Danube alluvium which contains little iron oxides. Forestry-steppe zone supplied less organic matter in surface layer. Both factors, small amount of iron oxides and organic matter, imply very low values of susceptibility in the whole profile; in upper horizons $\chi = 25 \times 10^{-8}\ m^3\ kg^{-1}$ and in parent rock $\chi = 10 \times 10^{-8}\ m^3\ kg^{-1}$. In SNO profile, also pedogenesis is not so well developed as in the remaining profiles.

Kastanozem ASK was formed in warm and dry regions with not so dense vegetation cover as on chernozem steppes. This specific vegetation type causes that more than 50% of all roots are concentrated in the upper 25 cm of the soil. Eurasian Kastanozems have Ah horizon of about 50 cm thick, but in the drier south (as in case of ASK profile), the Ah-horizon is only 25 cm thick. The occurrence of argic B horizon in Kastanozems demonstrated by increase of susceptibility is still ill understood. They may be fossil, as claimed by some Russian soil scientists (Eckmeier et al. 2007), but there are also theories of a more recent formation, through 'normal' translocation of clay, or by destruction of clay or fine earth near the surface and re-formation at greater depth. Kastanozem has a water regime with intermittent dry and wet periods which causes that weathering and pedogenic processes take place at shallow depth. Low thickness of humus layer is in good accordance with low thickness of magnetic enhancement layer. Independently of LOI high SP grains content in parent rock indicates chemical alteration of loess layer before soil started to form. Relatively large grain size in A horizon points to advanced pedogenesis and to maturity of soil.

According to Liu et al. (2013), χ_{fd}/HIRM slope is significantly correlated with precipitation. If we accept χ_{fd}/HIRM slope as a proxy for precipitation our results indicate that SEN, RUS and ASK profiles were formed in climate of approximately similar moisture whereas SNO was formed in much drier climate conditions. As present day precipitation values do not indicate such significant differences, low χ_{fd}/HIRM slope for SNO can be explained by a negative water balance. Importance of water balance for pedogenic processes was postulated by Orgeira et al. (2011). In the case of SNO it can be caused by high permeability of carbonate fluvial sediments which are the parent material of SNO soil.

6 Conclusions

1. In all the profiles, the first pathway of iron oxides formation is observed, that is, weathering of Fe-bearing minerals such as smectite to ferrihydrite, and then formation of a transient intermediate phase—hydromaghemite from ferrihydrite, followed by transformation of hydromaghemite to hematite and/or to goethite, depending on soil environment.

2. The ratio of χ_{fd}/HIRM allows to evaluate approximately similar precipitation for ASK, RUS and SEN and much lower for SNO. As SNO should be formed in the same conditions as RUS and SEN, we probably observed the effect of higher permeability of parent rock in SNO profile sediments than in the case of RUS (granite) and SEN and ASK (loess).
3. Relation of LOI to magnetic parameters indicates that organic matter promotes pedogenesis progress.
4. It is possible to relate vertical magnetic structure of soil to pedogenic differentiation of studied profiles.

Acknowledgements This study was supported within statutory activities No 3841/E-41/S/2015 of the Ministry of Science and Higher Education of Poland. We are grateful to Dr. Igor Tunyi for his assistance in field work during sampling of soil profiles in Slovak Republic.

References

Barrón V, Torrent J (2002) Evidence for a simple pathway to maghemite in Earth and Mars soils. Geochim. Cosmochim. Acta 66:2801–2806

Canarache A, Vintila II, Munteanu I (2006) Elsevier's dictionary of soil science. Elsevier

Cornell R, Schwertmann U (2003) The iron oxides. Structure, properties, reactions, occurrence and uses, 2nd edn. Wiley VCH, GmbH&Co, KGaA

Day R, Fuller M, Schmidt A (1977) Hysteresis properties of titanomagnetite: grain size and composition dependence. Phys. Earth Planet. Inter. 13:260–267

Dearing J, Hey K, Baban S, Huddleston A, Wellington E, Loveland P (1996) Magnetic susceptibility of soil: an evaluation of conflicting theories using a national data set. Geophys. J. Int. 127:728–734

Driessen P, Deckers J, Spaargaren O, Nachtergaeleal F (2001) Lecture notes on the major soils of the world. World Soil Resources Reports. 94 FAO, Wageningen

Dunlop DJ (2002a) Theory and application of the Day plot (Mrs/Ms versus Hcr/Hc): 1. Theoretical curves and tests using titanomagnetite data. J. Geophys. Res. 107(B3):2056. doi:10.1029/2001JB000486

Dunlop DJ (2002b) Theory and application of the Day plot (Mrs/Ms versus Hcr/Hc): 2. Application to data for rocks, sediments, and soils. J. Geophys. Res. 107(B3):2057. doi:10.1029/2001JB000487

Eckmeier E, Gerlach R, Gehrt E, Schmidt MWI (2007) Pedogenesis of Chernozems in Central Europe—a review. Geoderma 139:288–299

Evans ME, Heller F (2003) Environmental magnetism: principles and applications of enviromagnetics. Elsevier Science, Academic Press, San Diego (USA)

Eyre JK (1997) Frequency dependence of magnetic susceptibility for population of singe-domain grains. Geophys. J. Int. 129:209–211

Fanning DS, Fanning MCB (1989) Soil morphology, genesis and classification. Wiley, New York

Hanesch M, Petersen N (1999) Magnetic properties of recent parabrown-earth from Southern Germany. Earth Planet. Sci. Lett. 169:85–97

Hanesch M, Scholger R (2005) The influence of soil type on the magnetic susceptibility measured throughout soil profiles. Geophys. J. Int. 161:50–56

Head KH (1992) Manual of soil laboratory testing, vol.1. Pentech Press, London

Heiri O, Lotter AF, Lemcke G (2001) Loss on ignition as a method for estimating organic and carbonate content in sediments: reproducibility and comparability of results. J. Paleolim. 25:101–110

Hoogsteen MJJ, Lantinga EA, Bakker EJ, Groot JCJ, Tittonell PA (2015) Estimating soil organic carbon through loss on ignition: effects of ignition conditions and structural water loss. Eur. J. Soil Sci. doi:10.1111/ejss.12224

Hrouda F (2011) Models of frequency-dependent susceptibility of rocks and soils revisited and broadened. Geophys. J. Int. 187:1256–1269

Jeleńska M, Hasso-Agopsowicz A, Kopcewicz B, Sukhorada A, Tyamina K, Kądziałko-Hofmokl M, Matviishina Z (2004) Magnetic properties of the profiles of polluted and non-polluted soils. A study case from Ukraine. Geophys. J. Int. 159: 104–116

Jeleńska M, Hasso-Agopsowicz A, Kadzialko-Hofmokl M, Kopcewicz B, Sukhorada A, Bondar K, Matviishina Z (2008a) Magnetic structure of the polluted soil profiles from eastern Ukraine. Acta Geoph. 56:1043–1064

Jeleńska M, Hasso-Agopsowicz A, Kądziałko-Hofmokl M, Sukhorada A, Bondar K, Matviishina Zh (2008b) Magnetic iron oxides occurring in chernozem soil from Ukraine and Poland as indicators of pedogenic processes. Stud. Geophys. Geod. 52:255–270

Jeleńska M, Hasso-Agopsowicz A, Kopcewicz B (2010) Thermally induced transformation of magnetic minerals in soil based on rock magnetic study and Mössbauer analysis. Phys. Earth Planet. Inter. 179:164–177

Jenny H (1941) Factors of soil formation. McGraw-Hill

Jordanova D, Jordanova N (1999) Magnetic characteristics of different soil types from Bulgaria. Stud. Geophys. Geod. 43:303–318

Jordanova D, Jordanova N, Petrov P, Tsacheva T (2010) Soil development of three Chernozem-like profiles from North Bulgaria revealed by magnetic studies. CATENA 83:158–169

Kądziałko-Hofmokl M, Kruczyk J (1976) Complete and partial self-reversal of natural remanent magnetization of basaltic rocks from Lower Silesia, Poland. Pure Appl. Geophys. 110:2031–2040

King J, Banerjee SK, Marvin J, Özdemir Ö (1982) A comparison of different magnetic methods for determining the relative grain size of magnetite in natural materials: some results from lake sediments. Earth Planet. Sci. Lett. 59:404–419

Kruiver PP, Dekkers MJ, Heslop D (2001) Quantification of magnetic coercivity components by the analysis of acquisition curves of isothermal remanent magnetization. Earth Planet. Sci. Lett. 189:269–276

Liu QS, Banerjee SK, Jackson MJ, Maher BA, Pan YX, Zhu RX, Deng C, Chen FH (2004) Grainsizes of susceptibility and anhysterec remanent magnetization carriers in Chinese loess/paleosol sequences. J. Geophys. Res. 109:B03101

Liu QS, Deng C, Yu Y, Torrent J, Jackson MJ, Banerjee SK, Zhu RX (2005) Temperature dependence of magnetic susceptibility in an argon environment: implications for pedogenensis of Chinese loess/palaeosols. Geophys. J. Int. 161:102–112

Liu QS, Barrón V, Torrent J, Eeckhout SG, Deng LC (2008) The magnetism of intermediate hydromaghemite in the transformation of 2-line ferrihydrite into hematite and its environmental implications. J. Geophys. Res. 113:B01103

Liu Z, Liu QS, Torrent J, Barrón V, Hu P (2013) Testing the magnetic proxy χ_{FD}/HIRM for quantifying paleoprecipitation in modern soil profiles from Shaanxi Province, China. Global Planet. Change 110:368–378

Maher BA (1986) Characterization of soils by mineral magnetic measurements, physics of the Planetary Interiors, vol 42. Elsevier Science Publishers B.V, Amsterdam, pp 76–92

Maher BA (1998) Magnetic properties of modern soil and quaternary loessic paleosols: paleoclimatic implications. Palaeogeogr. Palaeoclimatol. Palaeoecol. 137:25–54

Maher BA, Thompson R (1995) Paleorainfall reconstruction from pedogenic magnetic susceptibility variations in the Chinese loess and paleosols. Quaternary Res. 44:383–391.

Maher BA, Thompson R (eds) (1999) Quaternary climates, environments and magnetism. Cambridge University Press, Cambridge, UK

Maher BA, Thompson R, Zhou LP (1994) Spatial and temporal reconstructions of changes in the Asian paleomonsoon: a new mineral magnetic approach. Earth Planet. Sci. Lett. 125:461–471

Maher BA, Alekseev A, Alekseeva T (2003) Magnetic mineralogy of soils across the Russian Steppe: climatic dependence of pedogenic magnetite formation. Palaeogeogr. Palaeoclimatol. Palaeoecol. 201:321–341

Mullins CE (1977) Magnetic susceptibility of the soil and its significance in soil science: a review. J. Soil. Sci. 28:223–246

Orgeira MJ, Egli R, Compagnucci RH (2011) A quantitative model of magnetic enhancement in loessic soils. In: Petrovsky E, Ivers D, Harinarayana T, Herrero-Bervera E (eds.) The Earth's magnetic interior. IAGA Special Sopron Book Series, vol. 1. Springer, Heildelberg, pp 361–397. ISBN:978-94-007-0322-3

Santisteban JI, Mediavilla R, López-Pamo E, Dabrio CJ, Zapata MBR, García MJG, Castaño S, Martınez-Alfaro PE (2004) Loss on ignition: a qualitative or quantitative method for organic matter and carbonate mineral content in sediments. J. Paleolim. 32:287–299

Schwertmann U (1985) The effect of pedogenic environments on iron oxide minerals. Adv. Soil Sci. 1:172–200

Schwertmann U (1988) Occurrence and formation of iron in various pedoenvironments. In: Stucki JW, Goodman BGA, Schwertmann U (eds.) Iron in soil and clay minerals. Reidel Publ. Co. Dordrecht, Holland, NATO ASI Series, vol. 217, pp 267–302

Thompson R, Oldfield F (1986) Environmental magnetism. Allen & Unwin, London, UK

Torrent J, Liu QS, Barrón V (2010) Magnetic minerals in Calcic Luvisols (Chromic) developed in a warmMediterranean region of Spain: origin and paleoenvironmental significance. Geoderma 154:465–472

Van Velzen AJ, Dekkers MJ (1999) Low-temperature oxidation of magnetite in loess-paleosol sequences:a correction of rock magnetic parameters. Stud. Geophys. Geod. 43:357–375

Worm HU (1998) On the superparamegnetic—stable single domain transition for magnetite, and frequency dependence of susceptibility. Geophys. J. Int. 133:201–206

Kolibri 360 GmbH
Ferdinand-Jühlke-Straße 7
99095 Erfurt, Deutschland
produktsicherheit@kolibri360.de